# KILLER TECH

## TECH

### AND THE
### DRIVE TO SAVE
### HUMANITY

# MARK STROSS

**KA**

**KATHMARK.**
**PUBLISHING**

Published by Kathmark Publishing–Houston, Texas

Cover Photo Credit: Colt Melrose Photography

Cover Design Credit: Silversmith Press Creative

Printed in the United States.

ISBN: 979-8-218-97037-6 (Hardcover Book)
ISBN: 979-8-218-97038-3 (Softcover Book)
ISBN: 979-8-218-97039-0 (eBook)
ISBN: 979-8-218-97040-6 (Audio Book)

To all of my life's teachers.

To all of my life's teachers.

# Contents

*Foreword* ........................................................ix

*Preface: Backing It Up* .................................... xiii

*Acknowledgments* ........................................ xxi

**Chapter 1:**
Human Productification.................................... 5

**Chapter 2:**
Life After Bandwidth......................................... 23

**Chapter 3:**
The Hacking of Critical Infrastructure ................ 37

**Chapter 4:**
The Power of Being Informed (and Misinformed).............................. 42

**Chapter 5:**
The Unseen Cost of Technology ........................ 57

**Chapter 6:**
Cyber Warfare in Your Kitchen ........................... 72

**Chapter 7:**
Reality vs. Virtual ............................................ 87

**Chapter 8:**

The Illusion of Ownership ................................................... 109

**Chapter 9:**

The Illusion of Power — Green Power ................................. 121

**Chapter 10:**

A Special Kind of Courage ............................................... 131

**CHAPTER 11:**

Your Digital Hygiene ........................................................ 145

**CHAPTER 12:**

Bringing Humanity Back .................................................. 155

*The Making of Killer Tech: Mark's Back Story* .................................... 177

# Foreword

By Donetta Colboch

I've known Mark since 1991 when the Internet wasn't a thing, a cell phone was the size of a brick (if you had one), my work computer was a tiny Macintosh, black and white with a monitor built-in, cable had only a few channels and MTV was nothing but music videos. Mark and I met at National Association of Broadcasters (NAB) in April of that year. I was a new member of the marketing team for NewTek and Mark was one of our early and most spirited customers.

As I look back on my history with Mark, what stands out to me is the way he enthusiastically consumed and processed information unlike anyone else I had ever met. Mark spoke so fast and had so many ideas running through his head it was like listening to a tornado! He chewed through information like a chainsaw and devoured everything he saw and heard—like the Looney Tunes Tasmanian Devil! One of my favorite memories was taking Mark to lunch in LA with a friend from a local university in Topeka where NewTek was founded. The university was creating a training program based on NewTek technology and wanted to hear what Los Angeles' experts using our technology would find valuable for the students in school. At the end of the lunch when we parted ways, my friend turned to me and said, "That was amazing! I learned so much, but do you realize, he never stopped talking? Yet, when I looked down, his meal was gone without taking a break or spilling a bite!"

I watched over the years as Mark took on increasingly complex projects; he always worked on the bleeding edge of technology. He would find a new technology, NewTek's, or others, and push it to the limit. One example, NewTek had a new product in development for a lengthy time and we were under pressure to get it completed and on the market. NewTek's founder asked me whom I would suggest who would be interested in buying the first unit, knowing that it worked, but it was untested and likely difficult to implement. Of course, the first person I mentioned was Mark Stross! His reaction was, "Absolutely, if anyone could make this work it's Mark." We were on a deadline, so our company founder flew from Topeka to the Kansas City airport by helicopter to the FedEx tarmac to deliver the system. When the system arrived, Mark took it to Cleveland where he had the system up, functioning and displaying graphics on the big screen in Cleveland Square the next night!

Year after year, I watched Mark as he sought out new technologies, never as the developer or the programmer, but as a visionary who sees beyond the brochures and the spec lists, to the potential others overlook. Mark has grown far beyond the whirlwind of ideas I met in the 90's to a seasoned professional who understands, implements, and creates extraordinary visual experiences in the most iconic locations in the world.

Mark has a passion for the advancement of technology that has never wavered in the decades I have known him. He sees a future filled with great potential and dangerous pitfalls as our technology advances forward. I'm confident his passion will be as infectious to you as it has been to me and everyone I've ever known who has worked with Mark.

Whether you agree with everything you read here, you are wise to take note. Mark has an uncanny ability to imagine the future, and he's looking out for all of us, and he's looking to all of us to help make it "killer" in the best sense!

# Preface

## Backing It Up

Whenever I am asked about how I became the technologist I am today, I find the answer is really a simple one. Despite having a unique back-story, that includes parents with jobs that took us around the world, being an immigrant to America, and having found myself in countless serendipitous situations in some crazy places—my career and my life's work are the result of facing considerable odds and the relentless pursuit of progress.

At my core, I am an innovator who leads with a heart brimming with passion, all the while accompanied by an artist's fragile ego. In the early days, my desire for an immediate realization of my ideas, was often overwhelming to those who couldn't see what I envisioned. It was when I began to move my ideas from the drawing board to real life applications like the cutting edge technology at the World Trade Center or filling the ceilings of Penn Station with digital images of ethereal clouds, that my passion and drive started to resonate differently with others—through their admiration and inspiration, I found affirmation.

Throughout my journey, the need to confront and overcome fear has been constant. My way of doing things often included intense sparks of inspiration with passion, segueing into a quick execution, which often times led to spectacular failures. Ready, fire, aim. Yet, setbacks, are another part of the journey, and I have embraced them. I instinctively knew that failure is merely a stepping stone in the process of success. Through trials in my personal and professional worlds, challenges in marriage, family, my business relationships, and ever-changing technology; I've experienced the extremes of failure and success—the latter exceeding even my most wild imaginations.

Life with my amazing wife, Kathleen, is a tapestry of moments—some joyous, some challenging, and a few that are truly incredible—I honestly couldn't have made them up.

I'd like to share with you some of the defining moments of my life that have taught me the greatest lessons and led me to this moment; where you and I are connecting through the pages of this book.

From the outside, my upbringing would seem somewhat idyllic, although nothing is ever what it seems. My parents were in the film industry—my mother, a well-known British actress, and my father, a successful film producer. Because of this, I was often in the presence of great people with amazing stories—influential icons, renowned statesmen, and diplomats. Those stories could fill a book of their own. Still, as it relates to this one and how I got to where I am today as a technologist, the underpinning of my success is that these highly influential people influenced me—with their words, actions, and experiences.

During my childhood in Switzerland, I had the extraordinary privilege of sharing regular lunches with none other than Charlie Chaplin, who was in his 80s at the time. This cinematic legend, a wonderful and wise man, took it upon himself to impart invaluable life lessons to me. One particular encounter with Charlie left an indelible mark on my young mind. He asked me to close my eyes and envision him in his iconic role from the 1930s silent film, *Modern Times*. In this classic, Charlie Chaplin portrays a factory worker engulfed by the machinery he's trying to control.

As a wide-eyed boy, it was challenging to reconcile the image of the elderly man before me with the dynamic actor I had seen on TV. Sensing my skepticism, Charlie turned this moment into a profound lesson. With eyes shut tight, I imagined him navigating the cogs and gears, mirroring his character's struggle in the film. When I opened my eyes, he declared, "You see; I did." In that transformative instant, Charlie unveiled the immense power of my imagination and its profound impact on my belief. Developing this power of my imagination would prove to be one of my greatest assets.

Another pivotal chapter unfolded when I was around 14, when I had the remarkable opportunity to meet Mohammad Reza Pahlavi, the Shah of Iran. At the time, Iran was a secular monarchy, heavily influenced by Western and British ideals. The Shah championed women's rights, education, land ownership, and forbade traditional Islamic attire. A fan of my mother's work, the Shah extended an invitation for our family to spend Christmas with him—a rather unconventional gesture for a nation's leader.

Imagine, an opulent residence atop a snowy mountaintop setting, and an eclectic gathering of his family, government officials, and my parents and me. Amidst this surreal backdrop, I found myself seated at the designated children's table with the Shah's daughters; the princesses of Iran. We reveled in innocent childhood play, blissfully unaware of the impending seismic shifts in global politics that would redefine the significance of this memory.

The next year, the Shah's reign was abruptly terminated by Ayatollah

Khomeini, marking the onset of the Iranian Revolution and the rein-statement of traditional Islamic laws. The news reverberated worldwide, leaving an indelible imprint on me. It was surreal to witness someone I had personally known, the Shah, stripped of power and exiled with his family from their homeland. This profound experience taught me two important lessons. First, the transience of power and second, the abrupt, transformative nature of historical moments. Indeed, moments are undeniably real and hold the power to reshape the course of nations and individuals alike.

The fabric of my upbringing was further enriched by a great deal of adversity due to the choice of projects and subject matter my parents bravely took on. They created award-winning, transformational films that were highly controversial. My parents taught me how to go against the grain and face fear and uncertainty head-on. I did not understand at the time how much of an influence their bold behavior had on me, but it empowered me to embrace challenges and innovate fearlessly. I learned that failure is not the opposite of success; *it's part of it*. Not only did I learn how to "fail well" but how to also to get back up again, and again, and again.

I have always believed everyone possesses a superpower or some-thing they are uniquely equipped to do. As a younger man, I would have said that my superpower was to take existing technologies, modify them, and find new markets and uses. Today, I believe my greatest strength is possessing the boldness to stand against opposition.

In my career, I've been at the vanguard of creating "killer" technol-ogy—not in the ominous sense, but in the sense of what's new and cut-ting-edge. My innovations have steadily and often quietly become part of our everyday culture, and their effects are sometimes unbeknown to me.

Not long ago, my daughter, fresh out of Baylor University and embarking on her first business trip, reached out to me with a request. She wanted a recommendation for a good business book to read on the plane. Something that would grasp the essence of modern business. I advised her to pick any current business bestseller and read it cover to cover, so that she could understand the importance of knowing the language of modern business and thus, engage effectively in professional discussions.

Several hours later, she excitedly called me back, and told me she had purchased the New York Times bestseller, *The Latte Factor*. She stumbled upon a captivating story in the first chapter—an account of the author's daily passage through the Oculus in New York City. The author emphasized the significance of intentionally absorbing the magnificence of the world and not missing out on the splendor surrounding us.

For those unfamiliar with the Oculus, it stands as a remarkable transit hub, replacing the former Port Authority Trans-Hudson station which was destroyed during the 9/11 attacks. This stunning structure, resembling a dove released from a child's hand, serves as a $4 billion dollar memorial to those who fell that day and the testament to resilience which followed. Adorned with countless interconnected video boards, it displays a profound story with images. As the Chief Technology Officer at ANC, the company responsible for the installation; I crafted the software to operate the video boards.

When your daughter stumbles upon the opening tale of a New York Times bestseller—one that mirrors the impact of your work, it's a moment of realization. My technology had become woven into the city's fabric and into its culture. It was a profound reminder of the privilege of having contributed to something meaningful.

In this book, my aim is to foster a different kind of cultural impact—to educate and empower individuals worldwide, enabling them to embrace technology without becoming a slave to it or succumbing to the dominance of the corporations that create it. I want you to harness the positive aspects of technology without compromising your essence.

Technology is undermining our creativity, critical thinking, and the very essence of our humanity. As advancements in technology persist, we have been reduced to profit-generating entities, threatening what our intrinsic worth—the part that makes us human. We're treated as simply "users," dehumanized into numbers, devoid of our unique identities.

Being human means valuing and respecting others while employing critical thinking to make informed choices, not succumbing to programmed responses like Pavlov's dogs. Our humanity is eroding before our eyes, but it's not too late to act. This is a cause worth fighting for. Wouldn't you agree?

You might wonder how we embark on this battle. Just asking that question is your first step. Transformation begins with curiosity, awareness, and then action. We face the same challenges that generations before us encountered, albeit in a new form—no longer muskets and cannons, but rather algorithms and machines, powered by greed.

I want you to recognize your power. You are more influential than you realize. As you go through this book and learn the steps to take to safeguard your rights, you will be equipped to drive change at the local, state, national, and global levels. You will ignite a transformative moment for yourself, your family, your community, and future generations.

Start by saying "yes" to making a difference. Now is the time to stand up and reclaim the technology conversation. This book is about the drive to save humanity, and I challenge you to join me.

.

# Acknowledgments

What is a dyslexic technologist doing writing a book? It comes from passion and people helping me achieve my full potential, even when I didn't believe in myself. I have been lucky to have traveled and experienced the world as I was growing up. My parents, being in the entertainment industry, exposed me to experiences and ideas that have helped me become who I am today. I am grateful for the lessons I learned about being successful and to never to give up. They taught me: if you believe in your art, fight for it!

So, this acknowledgment goes to the people who helped inspire this book:

**I begin with the love of my life, Kathleen**: You complete my life and taught me many lessons on becoming a father and husband. I am grateful for the help and love you offer me daily. You always give me space to take risks, to try, and you support the outcomes. It is a privilege to be married to you.

**Ryan, Maddie, Angelica, and Spencer**: Thank you for being outstanding representatives of our family! Mom and I took two families and merged them to form us—a family without steps. You've inspired me to write this book to help protect your futures. I am so proud of you all, individually and in unique ways.

**Joanna Hunt**: A coach, an author, a publisher, a fierce fighter for our rights; and a friend who told me to write this book and that she would have my back. Joanna you graciously agreed to edit and write with me. Everything you told me transpired—including all the different levels of hell you go through mentally writing a book! (Amazingly, you become your own judge, jury, and executioner, all in your head.) Thanks, Joanna, for getting me out of my head and committing words to paper.

**To Jerry Cifarelli Sr.**: This book was not a planned event in my life, but your simple "encouragement" (well…order) to reach out to Bulldog after debuting my music albums on his radio show was the seed that started it all. Thank you for your friendship and for taking a risk on a much younger version of me 25 years ago. Over the years, you've been my boss, my friend, at times my antagonist, other times my mentor and always a loyal business partner. Jerry, thank you for believing in me more than I did myself at times. Twenty-five years later, the second generation has taken over ANC—the people we trained together. Our ability to stay together all these years built a great business and created countless great stories to talk about! What an adventure—creating world-record setting technologies together. Your leadership inspired dreams to manifest.

**To Bulldog**: It was an honor to be interviewed on the air by you the first time, and even more so that it evolved into our weekly radio program together, "*Tech Byte*: *The Edge in Everyday Technology*." Can you believe that as of this writing, we've done 170 shows together? It's been much fun, and I have enjoyed meeting many wonderful people. Marla, Dude, and Coach, and everyone who is a part of the Rude Awakening show and the staff of Ocean 98.1 in Ocean City, MD—a massive shout-out for helping me "the newbie" on the air. I'm so grateful for this opportunity.

The research and weekly discussions on this show inspired the body of work for this book!

**To Donetta Colboch**: You have known me over 30 years! How blessed I was to meet you when you were doing marketing and PR for NewTek. You have seen my many failures and successes, always being there to help me get back up on my feet—or keep my ego in check—many thanks for being a friend and business ally for 3 decades.

**To Jerry Cifarelli Jr.**: Working with you and watching you pick up the baton to lead ANC has been a blast! Thank you for believing in me and allowing me to continue the journey under your leadership. Jerry, I'll never forget our trip to Napa Valley 5 years ago and the proclamations you made about your life and destiny. Here we are today and I've become a part of the reality you proclaimed those years ago—which is a bit uncanny and a testament to your vision and determination. I appreciate you and I appreciate working with you.

**To Kevin Ross, Jeff Paulson, David Compton, Mark Eggenberger, and Jackson Hart**: The software technology team at ANC is one of my proudest achievements! Thank you from the bottom of my heart. You are the definition of a dream team and consummate professionals! You make technology dreams happen and you've taught me many important lessons.

**To the entire staff at ANC**: I'm so proud to work with each of you. I couldn't ask for better coworkers to bring the vision to life.

**Dr. Steve Turley**: You don't even know that you first taught me about the network vs. mass society concept on your podcast, Turley Talks. This inspired me to dig deeper and address the concept in this book. Thank you! I always enjoy your patriotic work for Americans and hope to work closely with you to help all Americans stay safe around AI and technologies.

**Kim Kelley**: Thank you for inspiring and believing in me. You continue to show up and you are a great friend. Your work with Veterans for Child Rescue is inspiring and so meaningful. Thank you for getting child traffickers and sexual predators off the street! Everyone at V4CR, you're doing great work! Looking forward to doing more with you all.

A few special friends have helped me during the many issues that came up writing this book: Sheree Morgan, thank you for your family advice. Chris Boyer, thank you for believing in me enough to do the Discovery Program. To everyone involved with the Discovery Program in Austin, TX—you changed my life and helped so many people around me. Your program works! Thank you also to Dr. Penchas, and all the staff at Midtown Dentistry for giving me a new smile. When I smile, people respond differently. Thank you! Shout out to Christy DiGiambattista, who has always believed in my potential and has opened doors and my mind to what is possible. Thank you to the esteemed mayor of Ocean City, Rick Meehan for your warm welcome and for bantering with me and the Bulldog team on air over technology stuff that you thought you didn't need to know!

And in closing, while writing this book, music rescued me from distractions and allowed me to get in my zone. The albums I listened to the most were from the artist called "Spunkshine" and his collection of thinking music. Spunkshine, I'm grateful for your talent! And thank you to *Twining's English Breakfast Tea*, which I discovered is better for writing than coffee. But maybe that's because I'm British.

Finally, to you the reader: I'm grateful you are holding this book in your hand. I hope it draws out your humanity and inspires you to take action for yourself and future generations.

# KILLER
# TECH

# KILLER
# TECH

# CHAPTER 1

# Human Productification

*If you're not paying for the product;*
*YOU are the product.*

What image comes to mind when you hear someone described as a "user"? The connotations aren't pretty. In the drug world, users are addicts. To a dealer, users are just a number, a means to an end. They generate revenue. There's a reason that tech platforms call people "users." There's a reason they call *you* a "user." Referring to someone as a "user" removes their humanity, thereby making it easier to *use* them. In short, users use and get used. Humans on the other hand, have value. Humans dictate experiences and have rights. They have ideas. They have something to say and want to be heard, valued, and respected. At least, they should. We have unalienable rights given to us by our Creator, which should be acknowledged and transcend all platforms and spaces. I'll get into this topic more in just a little bit. The problem is that most people don't even know how close we are as a society to falling off the edge of technology into a world where Artificial Intelligence (AI) outpaces and outperforms human touch in every aspect of life. Many humans behave like "users" consuming technology like junkies, largely unaware they are being manipulated for profits. Those who *are* aware have no idea what to do about it. But by the time you finish this book, you will be both aware and empowered. You already have more power than you realize. However, you must understand what I am going to lay out for you, and you must be committed to protecting yourself, your family, and future generations. Buckle up, dear friend. The ride I took with technology in the first part of my life is nothing compared to the ride we will take together.

## The Pace of Change

The pace of change today is so fast and furious that most people find it easier to tune out than to understand the tech world around us. But if we tune out, we miss out—on a lot. As we explore the tremendous potential, pitfalls, and very real risks of today's technology you'll see that technology creates incredible opportunities, but yet we need guardrails to keep us from falling off the "digital cliff." I confess I am addicted to modern culture, and you probably are, too. The technology, opportunities, and infrastructure of our era are the greatest of any time in history. The question is, how much of your humanity are you willing to give up for the sake of convenience and free stuff? Do you even recognize *what* you are giving up? Most people don't. Think about this: If you're not paying for the product; *YOU* are the product, and whoever is selling you, owns you. The companies that are selling your digital data, effectively own your digital presence.

You may not view yourself as a product, but I assure you, Big Tech, marketing companies, the media, and even the government do. Algorithms monitor your habits and then place you into special interest categories and sell your eyeballs to the highest bidder—if you let them.

And if you aren't proactively stopping them, you are allowing them. Here's how it works: Because there is a limited number of categories that can exist, people are funneled into a finite number of existing categories based on the stuff we like and the content we consume. Machine learning further subdivides us into funnels of censorship and propaganda to help sway us into a desired action or outcome. In other words, despite your uniqueness, the online world reduces and categorizes you to a formula, where you, the "user" are auctioned off as a commodity. Dehumanizing, isn't it?

Think about this: *If you are being studied and fed information that will influence your behavior, how free are you in the online space?* If you think you have choices, but your selections are carefully curated, are you really choosing and exercising free will? If not, who is in control? Well, I'm glad you asked. The answer is the tech giants, the corporations creating the programming and narrow AI rules. This needs to change. And how is all this perceived choice impacting people? It's not good. Worldwide discontent has increased as social media becomes more powerful and dehumanizing. Problems are being created that were never problems before such AI rules existed. Because of the algorithms that curate social media feeds, billions of people are conditioned to live through the lens of their own bias, reinforced by the stream of information they are served. *Mind control, anyone?* If you want to travel down an interesting rabbit hole, look up "Operation Mockingbird"—*after* you finish this book. At the same time, in countries that have shielded themselves from the propaganda, like Hungary, people live remarkably happier lives. So, it doesn't have to be this way in America or any other country either. Depression and anxiety can be mitigated by simply putting the phone down. This takes discipline but it means you stop allowing products to define the way you see yourself or your worth. Humans—we have more power than we know!

So, when you think of your online and digital experiences, understand that there are two highways to travel: one that is free, where you are the product and content is carefully curated for you in a way that influences your behaviors, and another that is "pay-as-you-go" like a toll road. Option two is more exclusive with less traffic, and the pièce de résistance is that it's your own "freeway" of your design, not someone else's design for you. We must never forget that human beings are not the sum of their bank accounts, social security numbers, or the value of their likes and followers. Instead, we are a collection of our life stories and experiences that comprise who we are at our core, and we deserve to be seen as unique individuals and express who we are authentically. "Users" are addicts; owned, controlled, and treated as commodities. Humans are unique, free thinkers who will fight for their free will. Which one are you?

By the end of this book, I hope you no longer identify as a "user" and that, instead, you graduate back to being a human with a soul, mind, choice, and power.

## Genned Up Junkies

If we are going to return to being human and hold that position, we must take a good, hard look at ourselves in the mirror and get honest about how we got here. As it is today, the truth is, "We are all genned up." When I first said this to my wife, Kathleen, not long after getting started on this book, she looked at me peculiarly. (Which is not uncommon around our house.)

"Genned up?" she replied.

"Yes, genned up."

"What are you talking about?"

I realized my British roots were showing, albeit apropos. "Genned up" is a British saying I learned growing up in Europe, and it very accurately describes our global culture today. To put it in plain English (British-English anyway) "genned up" means that a person has found out as much information as possible and is *overloaded*.[1] Fits perfectly, wouldn't you agree? Friend, our culture at large is on complete information overload. And to make matters worse, not only are we in information overload, but we are also addicted to it.

What's the best way to create a loyal following? Hit their dopamine centers and make them junkies, craving more, like "users." We now have an entire generation of people raised as passive content consumers, addicted to their tech; the outcomes of which we do not yet fully understand. We haven't even begun to see the ramifications of this massive cultural shift.

Everything digital has consequences. Everything. You cannot predict the outcome, consequence, or impact of anything you post because an AI algorithm decides it. These outcomes are evolving constantly, and we need to pay attention and use discernment to understand what we are dealing with.

I learned this the hard way because I had a perspective that most parents have today. My biggest technological mistake was to think I could teach my children "common-sense" usage of technology, but how can an adolescent brain stand up against such a powerfully addicting, dopamine-stimulating activity? There's a reason some substances have age

---

1    https://dictionary.cambridge.org/dictionary/english/genned-up

limits, alcohol, cigarettes, etc.—because of the impact they have on the developing brain. Children can't make common-sense decisions about some things. As a father, I was wrong in thinking, *my kids need to learn technology early, and if they don't learn early, they won't be prepared for the future*. Wrong! The problem is that their young brains are too immature to grasp the difference between candy and poison. Sadly, the exact opposite of what I thought would happen became true. Handing technology to my kids too early and too often made it *more difficult* for them to handle the future.

Allow me please to jump on a very important soapbox. Parents—you have a difficult and often thankless job but remember we must be parents, not just friends to our kids. It's okay to say "no" especially regarding devices. Yes, they will beg, grovel, and try to wear you down, but stay strong for their own sake! Your job is to protect them, not just to please them. The child's brain is easily addicted because technology is designed to be addictive! It is designed to delight and engage and it does. Parents, we unknowingly created an entire generation of addicted "users," but there is hope. The path to a turnaround begins by turning it off. We must teach our kids to return to "using" their hands and brains to entertain themselves and solve life's problems. That's the way we were created! Never forget that neural pathways and coordination are developed by giving young bodies new challenges and experiences primarily outdoors and in nature. They will never get this in a digital world. No environment shown on a computer screen can stimulate growth and development the way the real world can. The controls used—the keyboard, mouse, or joystick—don't change. There's no new development by learning new computer games, no new hand-eye coordination restricted to a screen. If you don't get your kids exploring the real world,

the travesty is that they won't learn valuable life lessons and develop the tools they need for success in the future. Natural, unpredictable circumstances teach human beings resilience. Make sure your kids experience them! It is a fact that when a developing brain consumes too much screen time, pruning happens in the neural pathways. That's right; it's an actual dumbing down of our kids. How much is too much? The Academy of Pediatrics recommends no more than 2-3 hours a day of total screen time, including TV, phone, games, etc. Depression starts to develop at 5 hours.

Boredom is a gift to our kids. It is the space in which imagination thrives and creativity blossoms. Children today are not used to navigating physical things like boomers and Gen Xers did and they are not used to using their imaginations. They don't have to think. Instead, they have a phone that thinks for them. Today's kids don't have a chance to get bored. Many studies available online talk about how children are losing their imagination and creativity because electronic devices show them everything they need to know. Not only that, family interaction has radically changed too. For example, nowadays, families aren't playing chess and checkers like they used to. *Monopoly* boards, boardgames, puzzles, and card games sit collecting dust on the shelves. Now, the pandemic did fuel somewhat of a resurgence, but it's an important pastime. Playing games with your children is a way to teach them to use their imaginations and problem-solve. Not only that, boardgames are inexpensive compared to computer games. My dad used to play checkers with me, and he used to obliterate me. He never "let" me win. I hated that part, but you know what? I became better and more importantly, I had real conversations with my dad during those checker games. I got to speak to him about my future and how well I handled my life. I engaged with him on a level

I didn't get to during other activities. I also got to use my imagination. I would pretend that I had armies on the board and the armies would win. My point is that the whole experience was interactive and not digital, and it contributed to my overall development.

If you haven't done so lately, buy a boardgame! Buy something that will force you to engage with others and talk about valuable things like individual lives, friends, and current events. Be open to see things from a completely new perspective. Interacting with people is more important than consuming technologies. While we've drifted away from this important dynamic as a culture, we can bring it back. Starting today, do activities that promote human interaction, touch, and conversation. Community and family involvement is a powerful detox for our world.

• • •

Now, let's explore your relationship with social media, computer games, pornography, and other digital "entertainment." Get honest with yourself: Can you go a day without thinking about doing the thing you are addicted to? Probably not. Listen, no judgment. My guilty pleasures are YouTube and Rumble. I am a content junkie, guilty as charged, and I love to learn. Since I am a visual learner, these distribution methods appeal to me the most.

I mentioned my greatest technological mistake and now I have a story in which the protagonist is my youngest son. He has always loved computer games and has achieved remarkably high scores playing in different genres. From *Call of Duty* to *Minecraft*, my son loved to play hard and set records. I did not think I was sponsoring harm to my child, and I justified his gaming by telling myself that if I gave him access to digital tools, he would teach himself control and willpower. Of course, that was delusional thinking on my part. How could I put my son up against the combined research and development of massive tech companies that have spent hundreds of billions of dollars developing algorithms to make people addicted to their content? It wasn't a fair fight at all. I was wrong not to limit his usage time and access. Sadly, my son paid the price for my choice. In high school, he had no major passions or ambitions for the future. All he cared about was playing games. As a young adult, he dropped out of college in his first year and decided to work instead. He wanted to be able to play games on his laptop at will and live what he called a "simple life." He discovered that life isn't that simple, and entry-level jobs don't afford much luxury. While my wife and I were white-knuckling during this period, life taught him valuable lessons that no parenting could teach him. The first lesson he discovered for himself was that computer games left him empty and alone. His friends moved on with their lives, but he was not moving forward. He was largely alone, craving challenges that games could no longer satisfy. After realizing that gaming would not provide a fulfilling life, he came to me and Kathleen with a new plan. We were ecstatic! He eventually got on a new path and straightened out his life going forward. He's proudly serving in the US Army's 82nd Airborne Division; jumping out of perfectly good airplanes and helping protect our freedom. I could not be more proud of the man he's become.

Our story has a happy ending, but I wonder about kids and young adults who don't have a family network of support to help get them back on the straight and narrow. What will they do? Who will they be? Will they discover that life is more than technology, or will they tune out the small cry within that says, *Isn't there more to life?*

None of us are immune to the pull of the sophisticated algorithms that draw us in. Look at the millions of people who are using Facebook daily. I could ask any one of them why they spend so much time on this platform. Undoubtedly, they will tell me something like, "To keep in touch with family and get the news." So, I might say, "OK, can you stop using Facebook for one month?" Most people will say, "Of course I can do that." But then, when they try it, it's a different story. Statistics prove people break their resolution not to use a social media app within three days. Most people have no willpower against digital content—because, by design, we allow ourselves to become genned up!

The late Kenny Rogers left a wonderful bit of wisdom in the world— "You've got to know when to hold 'em, know when to fold 'em. Know when to walk away and when to run." In other words, you must know when it's time to call it quits and walk away—or run away—but it seems this advice has eclipsed our dopamine-addicted culture. "When to fold 'em" is not on anyone's mind. And that's what Big Tech is banking on—literally. They want your kids addicted, and they want you addicted because that drives profits and establishes control.

## Taking A Virtual Gamble

To add fodder to the fires of online addiction, traditional gambling in the virtual world is growing in ways none of us ever considered. Beyond that, we now have more creative ways to "take a gamble" in the digital world, such as buying and investing real money in virtual products and real estate. Buying real estate in the virtual world is a form of speculative investing, essentially gambling outside of the casino. If you invest in real estate in a metaverse, how do you know your investment will not be destroyed in a future update, rolling in like a virtual hurricane? You don't. And what if the business fails and the universe is turned off? Oh well…I guess you lost your investment. Say goodbye to your beautiful virtual apartment overlooking Snoop Dogg's palace. Of course, no one is banking on that going in. They hype is intoxicating as people focus on the potential upside, like a virtual gold rush. Technology may change, but people are still the same.

If you don't want to gamble in virtual real estate, you could try investing in virtual art or NFTs which is also a huge gamble when you step back and look at it. NFT stands for non-fungible token, like a digital receipt or proof of ownership in an online digital image or art piece. Yes, you can buy a piece of virtual art that exists only as code and proudly prove it is yours, even though there may be countless other versions online. NFT artwork is assigned a value and kept in a crypto ledger that everyone can access and see who owns it. It is a way to assign and prove the value of a virtual item. The real value of the NFT is in the eye of the beholders and believers, just like tangible art. Sort of. The difference is that with an NFT you are placing belief in a virtual, non-tactile item; the only proof of ownership is a digital key of zeros and ones—the coding language used to create it. One of the biggest controversies in the world of NFTs is how easily they can be copied—it is, after all, the digital world. While someone might pay millions for a particular NFT, others can screengrab it. Is it stealing? Sure. Does it happen? All the time. Not only that, but an artist can also create multiple copies of the same image as a series and sell them off. Sound a little risky to you? You're catching on. That is why NFTs are just another fad destined to be unseated soon. I go back to what Kenny said—"Know when to fold 'em…know when to walk away…and know when to run!"

At the end of the day, the biggest gamble we are taking in the online world is with our humanity as the lines between the virtual and the real are blurred.

## Beam Me Into Reality, Scotty!

As I mentioned, when I was growing up, my father was an accomplished film producer, and my mother was a well-known British actress. They traveled a lot and spent long hours on movie sets, which did not make time for a steady home life or leave much time for raising me. So, I was sent to boarding school in Monterey, California, to get a great education in a stable environment. I was around 10 or 11 years old, every weekday at 5:00 p.m. I would go to the common area at Robert Louis Stevenson School to watch *Star Trek* on the only public television on campus. I must tell you, that show taught me a lot about the type of person I wanted to grow up to be. I wanted to be like Captain Kirk, gallivanting around the galaxy. I give *Star Trek* much credit and kudos for shaping my young perspective. Some people grew up on *Star Trek*, others grew up on *Dukes of Hazard* and other well-written shows of the period. That is the beauty of well-produced narratives. They're impactful and last forever within the people who experience them. They influence culture in many ways. It's interesting to note that back in the 1960s, the crew members on board the Enterprise held digital pads that we now know as iPads. Many other *Star Trek* set items have become a reality, including doors that magically open as you walk up to them. Science fiction written in the past has become our present.

Here's an interesting story about *Star Trek* and confronting life's harsh realities. One summer, when I was at home in Beverly Hills, my dad told me, "Mark, I have a script, and I think it's good for William Shatner." He had invited William Shatner to our house to discuss the project. I couldn't believe it! Captain Kirk was coming to our house! The day he was to arrive, I was wide-eyed and filled with excitement all day long. Just before the time of the appointment, I was perched in our kitchen, peering out the window that overlooked our driveway. The buzzer from the driveway gate sounded, and I knew it was Captain Kirk. With my face practically pressed against the glass, I could see a little gold Honda Civic driving up to the house. The car parks, the door opens, and out pops Captain Kirk. But he doesn't look like Captain Kirk. Not at all. His clothes were different. His hair was different. There was a man walking up the pathway to our house, but it was not Captain Kirk—not by a long shot. At that moment, I had an existential problem, a glitch in my brain. I couldn't make sense of it. I expected to meet the Captain of the Enterprise in full uniform...but instead, I was staring at an ordinary guy in blue jeans.

This is just one example of how the make-believe world of digital entertainment fools our brains into thinking something is real when it isn't. Parents, we must insist that our kids leave the make-believe digital world and do a reality check. If not, their perceptions and expectations will be so skewed that when reality finally hits them, it will hit them hard...and square in the face!

## Thoughts About Living in a Simulation

In popular culture surrounding the Matrix movies, the idea of people being plugged into a simulation is gaining popularity. It's a known fact that Elon Musk believes we are not humans here on a real planet but that we are living in a digital simulation. So let's talk about that for a second—but only a second because it makes my brain hurt thinking about this idea. But for those interested, I'll give you the gist of where he is coming from.

Some people, like Musk, posit that we are drawn to virtual universes almost as if we know we are already in one. When we view the universe through the lens of our present understanding of technology, we can conclude as humans that the cosmos is just too big to store, and it requires a machine to make it all work; it's just too big for our brains to grasp. Along with that notion, when looking through the lens of a microscope, computer code and nature have similarities in their basic building blocks: zeros and ones. Symmetrical DNA strands correspond to coding metaphors. The more you drill down into both, the more similar they become. DNA has chemical switches like zeros and ones at its core, and computer code has zeros and ones at its core. That's how scientists are doing genetic splicing. They are reprogramming genetic code the way we would manipulate software code. It's a little bit awe-inspiring. There are magnificent similarities between nature and our digital universe. Entanglement theory suggests that two particles can be connected irrespective of location, which seems impossible.

However, there is an issue so profound that most glaze over it. How did everything in the universe start? Some say God, and some say nothing. But if you say "nothing," you have to ask yourself, *What created nothing?* Science can't answer every question, and so I guess a little faith is required either way.

## You Must Conform

What's even scarier than the blurred lines between the real world and virtual worlds, is that we are immersing ourselves into a virtual world that wants to program us to think, believe, and act according to their principles. I touched on this earlier, but did you know that Meta now has a program to "help" you if you feel that you're, in some ways, hooked up with an extremist organization? Yes, if you or your friend, whom you can report, has become a rebel because they were exposed to extreme ideas they want to help. The problem is that Meta defines what these extreme ideas are, and their definition of "extreme" includes ideas this country was founded on. Hmmm. Meta wants you to report your friends so they can seek "therapy" if their ideas don't line up with Meta's. Life After Hate is the name of the organization Meta developed to help change people's perspectives on "extremist" ideology, such as love of country, freedom of speech, and supporting our second amendment rights. It's a very interesting and potentially dangerous proposition that is now in the testing phase as of the writing of this book. Meta claims they have not decided if they will bring this out. The media covered it by saying, "This is a very good move because Meta is finally dealing with extrem-ists." The problem again is how do you define extremist? Is anyone who disagrees with your position considered a terrorist? That seems far out, but right now, people who love America, support individual liberties, and advocate for state's rights are considered "potential terrorists" according to government watch lists. So, there's that.

Few would argue that Meta should get an "F" for its bias. If Life After Hate is the organization they are sending people to, they are not trying to be unbiased. Now, there are extremes on both sides of the aisle. The question is, does Meta acknowledge this, or are they targeting one side of the aisle? Of course they are. It's no secret that Meta has biased policies with literally no governance towards the extreme left but all the governance towards the extreme right. I want to see governance on both sides and would like all extremes to be moderated fairly. If we cannot do things in an unbiased way, the house of cards will fold. And here's my prediction of how it will happen:

1. Meta will destroy itself and lose market share.
2. Meta will help divide the country further.
3. Censorship eventually kills its host.

I told you earlier I was raised by parents who embraced controversy, and I do too. I discussed this before the Twitter files were released on my radio show, *Tech Byte*. It  is important to understand how online social platforms run; promoting the ideology of those in power, in an effort to influence thinking and create change. These ideological factions are willing to rewrite and change outcomes. It's all driven by technologies, driven by algorithms, that resolve into zeros and ones. I don't want to go too far down the rabbit hole here because this book is not about politics or sides; it's intended to make you think like a human! Understand that everything today is driven by algorithms that are becoming increasingly complex and powerful. So, let me ask you: What does it mean for you to be a human?

In America, we have a document that defines and protects our unalienable rights as humans, the Bill of Rights, which is a part of our Constitution. But what about a bill of rights in the online world? We don't have that right now, but I believe we need a digital bill of rights that extends our human rights into virtual worlds. For example, do you have a say in your digital representation online? Why can't you own and control your data? What are the rules when a person dies? Who owns the digital content after a person is not around? Yes, there is a great need to create order and respect for your digital persona and assets—a digital constitution, if you will.

Technologies are dumb, but programmers are not. Their software represents their bias and the reality is that everyone is biased, whether intentionally or not. I don't believe in saints. Even people who start with good intentions evolve as power and wealth become apparent. This is why staying passive as consumers is not the answer. At this point, our proverbial backs face the water, and we have almost lost the beach! Let's not drown just because we love our free apps. We have not demanded much of Meta, X, Google, and other social platforms for the information they collect and the money they make on us. We've asked very little considering that when we sign in, we give up all our content rights and allow them to exploit our data any way they want according to their user agreement...which is subject to change. What the heck—if the agreement can be changed at the whim of the tech companies, then nothing is set in stone which means nothing can be trusted. This is why they want you addicted because you probably haven't even considered these things. It's time to wake up and take back our humanity.

## TECH ACTION

1. Have you ever looked at the usage logs on your phone or on your families' phones?
2. Do you have device usage limits for yourself and your family?
3. What's one thing you can do today to encourage more engagement with your loved ones offline?

Scan this QR code for additional information
on the topics in this chapter.

markstross.com/book-ch-1

# CHAPTER 2

# Life After Bandwidth

*We can't deny that technology impacts us,*
*both individually and collectively.*

We can't deny that technology impacts us both individually and collectively. It has changed how we interact, how our societies are organized, and how they are managed and governed. In short, we are moving from a mass society to a network society.

James Altucher had an intriguing feud with Seinfeld not long ago, and he was so upset that he wrote an op-ed about it. The feud was over the comment "New York is dead." I guess you should never pick on New Yorkers because New Yorkers take that very personally. Now, I'm not going to jump into the fray about who's right or wrong about New York because, frankly, that's not something I can litigate. However, I'm fascinated by the concept behind the article and something very interesting that Altucher brought up. In the article, he mentioned an intriguing concept: "Life before bandwidth and after bandwidth." Altucher's point was that the pandemic has forever changed how bandwidth has impacted culture.

The mass society is the product of the industrial revolution. For goods and services to be created, humans naturally gravitate towards geographic homogeny, manifested in creating cities. Humans found that massing up around products, human intellect, and brawn created the ingredients for societal evolution. It worked. Cities are the main driving force of all societies today. Everything major we do is announced and commenced within city boundaries, which means we are clumped together like sardines within city limits. The mass society worked best before the Internet, phones, and technology spread the news through proximity. Today, we don't need proximity; with modern technologies, transportation is no longer an issue. For the first time in human history, we don't have to be proximity-bound to be a part of that society. With Starlink now available, low-earth-orbit satellites change the game about where you can receive the Internet. Imagine you're living in Iceland in 2022. You will still get news within milliseconds of it being released. You can run a virtual office from Iceland, and no one will know where you are based, nor would they care as long as you deliver your service. Just 40 years ago, you could not run the same office. The expense of just the phone call from Iceland to the mainland would be prohibitive. Now, with the Internet, you can make a free Skype call and see the individual you're doing business with. Mass society is slowly dissolving away, and people are discovering new freedoms by doing business and life in creative ways.

The network society is the decentralization of people around cities. The impact is undeniable. I want to give a shout out to the man who introduced me to this concept, a man I have now met and have come to admire, Dr. Steve Turley. I knew nothing about the topics of the mass society and network society until I became a regular follower of his YouTube channel, Turley Talks, with well over 1 million subscribers. I admire people who start something new and grow it. With a hundred-dollar webcam and a consistent message, Dr. Turley has positioned himself as an American conservative authority. I respect that, and his information inspired me to dig deeper into areas of society I had never been exposed to.

Since the beginning of time, societies have been built on human capital, meaning that human intellect has powered the creation of everything we see, utilize, and consume. The only way for society to create and advance has been to group people together to collaborate and execute their creative ideas. We don't often think about this but imagine how useful a city is if you intend to bring people together into a dense area that requires many disciplines. For example, the car mechanic is close to the parts store, which is close to the dealerships; the restaurants, which are close to the dry cleaners, which are close to the chemical distributors and located around a geographic area to serve themselves better. Until now, it has been necessary to group people to communicate, transport goods, gather, share resources, and survive. However, today, in the "after-bandwidth era," we don't necessarily need these dense groupings of people anymore. Now, we can spread people out geographically and still create and accomplish goals through technology and modern transportation. And now, post-COVID, we have proven that corporations can still run without filling skyscrapers. It's astounding to see how productive our culture has been without everyone in a company returning full-time to the office. In addition, video conferencing, which used to consume a lot of bandwidth, can now be done in almost anyone's home throughout America, including rural areas. It's remarkable! Today, bandwidth is everywhere, and it's growing all the time.

Today, you can do business virtually with anyone, anywhere in the world, from anywhere in the world. Cities are transforming dramatically because of this shift. Businesses after COVID ask employees to return to the office less often, usually on a flex schedule of two days in the office and three out. This is a huge shift from the pre-COVID days for most employees.

When I started working at ANC 20 years ago, I proposed the idea of working remotely to my boss, Jerry, who was reluctant. At the time, I lived in Los Angeles and then moved to Houston. Jerry was used to a traditional business model where everyone reports to the office. I had to convince him to be open to something different, and I believed this was in the company's best interest. Not only for me, the Chief Technology Officer, but for my entire technology department. Working remotely allowed us to entice and keep some of the best programmers in the industry because we could offer them a unique lifestyle. It wasn't an easy transition to make. Setting up the equipment and the way we conducted the remote business was difficult. But, if the programmers came through on time every time, we didn't mind how they got their work done. We afforded them freedom, which was empowering and motivating for our team.

From there, we built ANC, and today, my programmers are located worldwide, and we have succeeded! This is my message to every businessman reading this book: The writing is on the wall—the old model of rented real estate in brick-and-mortar locations representing your brand is over. We are in the after-bandwidth era.

Understand that bandwidth is not the Internet, specifically; bandwidth is what is added to the Internet that allows the flow of information on the Internet. To explain further, bandwidth is a combination of many companies and governments working together to create the infrastructure for the Internet. It's everything from the fiber you put under the ocean to how you join it from country to country. This cooperative effort creates what we refer to as highways on which information travels. There is tremendous work done by each country to make sure they use all the bandwidth they build, and each country is constantly building new bandwidth, especially if they want to stay current. Then, with the advent of 5G—and soon to be upgraded to 6G, which will increase the speed of the flow of information—this requires larger "highways" to accommodate that bandwidth.

Now, understand that many pieces of equipment are used to create the World Wide Web. Hundreds of pieces of equipment may be between you and your friend on a Zoom call. Of all the equipment needed to make a connection, bandwidth is the weakest link and slowest part of the information highway. This means no matter how good your other equipment is, your signal is only as fast as your bandwidth. If you think of this infrastructure as a web—and we always say, "the World Wide Web"—but if you look at it as a concentrated web and look at each strand, how big is that strand? There's your bandwidth.

This probably isn't news, but we will never return to "the way it was." After bandwidth technologies exploded on the scene, society changed forever. Bandwidth allows a society to benefit from one another's intelligence (or stupidity). Bandwidth spread across a population allows that population to share a mass intelligence. People no longer must find a neighbor, a friend, or a local "expert" to ask a question. They go to the "experts" on the Internet. Why transact with a human if the world's intelligence is available in your hand?

## Connected Silos: Preference Censorship

One of the biggest challenges with migrating to a network society is how we are fed information, creating silos of thought. In business, a silo-mentality creates separate areas of expertise without the need or practice of sharing information between departments. This has proven to be damaging to the productivity of a business. Sharing information is essential for efficiency and creativity. In the network society, Google and other large tech companies are playing content curators to the world, creating silos of thought among the population. They're playing a balancing game between the algorithms they create to attract eyeballs to bring in advertising dollars, and the way they addict people to the advertising algorithm. They must feed people enough of what they want to keep them loyal and enough of what tech companies want to influence behavior. This means people are getting what Big Tech wants them to see, they aren't getting the whole story. The question becomes, *Are we crossing lines of morality and ethics with this model?* The tech companies always want to skirt the issue and say, "No, no, no, all we're doing is trying to make you stay on our platform. It's benign." The problem is that the algorithms have become so good at what they do. They're succeeding at manipulating humans and keeping us on their platforms. As discussed earlier, they do so by ensuring we get dopamine hits with content we want to see.

Unfortunately, humans will always gravitate toward the familiar and what makes them feel included or comfortable. Even if their comfort is in things considered on the edge, unpopular, or immoral, they'll gravitate towards seeing more of it. It has become apparent by observing the way the algorithms serve up fringe content, that people must gravitate towards it. So, what do you think the tech platforms are going to do to keep people on them? Yes, they will learn your bias and create more titillating fringe content that will keep you engaged. Think about that. A no-win situation will occur because in order for these empires to grow, they will keep delivering extreme content which fortifies people's silos of thought offline.

Now, I don't believe this was the original goal of the tech companies. I believe they had visions of each of their platforms growing. Still, I don't believe they could have envisioned what it meant for Facebook to acquire Instagram, for example, and all the other entities that Facebook has acquired. This gives the tech giant access to billions of people who subscribe to their services. Google, Apple, and many others are in the same boat. They have become huge. They have acquired so much influence. I don't believe any mortal human envisioned what they could have done with this capability. Should they be held accountable? Who should hold them accountable? This is a big problem, and breaking down the problem is simple: These companies are now responsible for so much of our daily digital lives and have very little accountability about how they conduct themselves. Their impact on society is huge, but their accountability is virtually nonexistent. And with so much influence and so little accountability, they have the potential to cause harm in a lot of different ways.

For example, in 2020, Google had two huge outages. The first on August 20, 2020, which was a global outage for several hours. The outage abruptly disrupted Google's suite of services, including Gmail, Google Drive, Google Docs, Google Meet, and Google Voice. On December 14, 2020, a second global outage occurred, affecting authenticated users of most Google services, including Gmail, YouTube, Google

Drive, Google Docs, Google Calendar, and Google Play. The problem was due to a failure in Google Accounts.

Why did this happen? It turns out that a routine update to the servers went wrong, and customers couldn't access their own content stored on those servers. This is scary. Why was it scary? Surprisingly, with so much of the world depending on them, they didn't have backup systems in place. I am not talking about having backup servers, but they should have had in place whole infrastructures running parallel so that if there was an issue, the secondary system would automatically take over. Or, even without redundancies in place, they could have performed updates regionally, instead of globally, so that only a small portion of customers would have been impacted instead of a worldwide outage. Seems like common sense, doesn't it? Unfortunately, Google seems to think it's okay to have only one set of authentication servers worldwide that can take out two billion customers when broken. These servers act as gatekeepers to people's digital lives, and it was amazing that Google had only one path. As a technologist, I was shocked at how vulnerable their authentication servers were! It's insane that a company that size isn't regionally separated when considering mishaps, war, or any general worldwide instability. If I were Google, I would have also created redundant server highways to ensure no one could hack into me or that a coding error couldn't break all my servers simultaneously. But Google somehow succeeded in taking themselves out across the whole world. And their attitude was, "It's not a big deal." In some circles, we call this gaslighting, but that's how well-programmed we are as the audience. Most people bought the idea. "Oh, well, it was just an outage." How many outages extend to two billion customers? In my world as a technologist in pro sports, ANY outage is a big deal!

Fast forward to the future; what would I like to see? I would like to see more accountability. It starts with you and me taking action on a local level. I would like to see that, starting right away, everyone becomes more engaged with their city leaders, right down to the school district level, which is a place where a person can make an impact simply by getting involved. You can change America by asserting your perspective and making your voice known to your local government.

Right now, the perspective that I think I'm most interested in asserting is that we must make sure technologies, especially in the classroom and in other places, don't become just one company representing the whole educational path. For example, Google took over the whole school district. To me, that is not good for our kids. We must ensure that we bring multiple perspectives into the classroom.

I believe all history needs to be taught. You can't just suddenly have one side of history; that's censorship. Ultimately, it's up to the people to decide what kind of government and technology they want. If Google and all the tech companies join up, and if our government, the last remaining counter to their strength, doesn't stand up to them, then at some point, they will be more powerful than any single government. Some people would argue they already have achieved that status.

By the way, I'm not against Google or Apple. I'm warning these companies. You are about to lose control of what you set out to be. When a tech platform decides one ideology is good and another is bad, and it creates bias in its search algorithms because that is their preference, isn't that preference censorship?

# The Power to Censor

In this book, we are talking about one consistent theme. This one theme is super important: the hubris and the extent to which these large tech companies will go to make money and influence people's behavior.

For example, it's hard to believe this, but did you know that Facebook unfriended a country? Yes. The entire country—Facebook unfriended Australia. Why? Because Australian media decided that Facebook should pay for using their original news content. Their reasoning was that if a platform, Facebook in this case, is going to censor politicians and free speech, it becomes a publisher; and if a platform is a publisher mone- tizing content, it should pay for the information coming into its servers. Facebook essentially said, "We don't like that." Google also wasn't going to play ball with Australia because of this, but they decided to start negotiating deals with Australia after about a week. Google agreed to incur the cost of content creation and paid the Australian media outlets.

Facebook, however, has the attitude, "We're so cool that by having your content on our platform, you will grow your country. Therefore, we should not pay for your country's content because we drive your growth. So, give us all your content for free." With that mantra, Facebook closed the whole of Australia down to having access to any news outlet, including their own, on it's platform. When they did, they took away access to two government medical sites dealing with COVID-19 and they took down access to all Australian government administration sites. By the end of the business day, they were forced to start bringing those sites back online. Many people criticized Facebook for restricting access to vital information and acting like it was bigger than the government of a country. Facebook returned services after reaching a deal with the Australian government to amend some aspects of the law. The deal gave Facebook more time to negotiate with news publishers and more discretion over which publishers it would support.

## Surveillance Capitalism

Some people are defining these tech companies in a very interesting way. The best way to describe these companies is to say they practice surveillance capitalism. Meta uses algorithms that tie into your computer, mobile apps on your pads, and all the apps on your phone. They are making money by tracking or surveilling your every movement online, and truly, offline as well. They did this until Apple and Google started to cut them off by limiting the amount of data they could get from people's mobile devices.

When Facebook (Meta) is downloaded onto your phone, it asks permission to connect to all your other apps and other things you do on your device, which means you're being tracked by Meta even when you leave Facebook. Apple is the first tech company to "declare war" against this. Apple has decided that they are going to brand themselves as privacy advocates. This is for real. In their operating system 14, Apple allows people to opt out of cross-tracking between apps. Google was upset by this but has agreed to the terms of it and is willing to play ball by allowing people to opt out of social tracking in apps. Meta said, "Heck no, we disagree and are taking Apple on." That's why Tim Cook and Zuckerberg have had some harsh words. You can watch all of this on YouTube. There are some great videos of the two tech giants posturing themselves around this idea that Meta wants access to the other apps on your phone without your permission. Apple says, "Hell no, you can no longer do this. This is unacceptable because the end user should know where their data goes." I believe Apple is in the right for what they're trying to do. I think Meta is really on the wrong side of history. I want to bring us all back to where it started.

As discussed in the last section, in Australia, Meta took away all the newsfeeds for the whole country to send a warning blow "across the bow" to all other countries and individuals saying that if you fight us, we will come back and harm you. Meta wants to convey that they are powerful—"You need us, and you can't live without us, so therefore, we're going to flex our muscles. Do it our way, or no way." This takes cyberbullying to a whole new level.

Interestingly, Canada and the EU have also decided to seek the same type of legislation Australia has. Countries standing up to Big Tech and protecting the rights of the people is the key to ending surveillance capitalism. Apple is including a request when you install Meta that says, "Do you want to be tracked? Yes or no?" And that has enraged Meta. They feel that this is an absolute assault on their freedoms and rights. However, what's amazing is that they put their rights and freedoms above ours! People ask me all the time if it's too late. Well, dramatically, it isn't. As people become informed about tracking, and begin opting out of it, businesses like Meta will be negatively impacted. Meta's tracking practice is already backfiring on them, as indicated by their stock price.

The bottom line is that Meta has acted like a teenager having a temper tantrum. A very first-world teenager with tons of privilege in a situation where they can't be sued, so they think they can do whatever they want. Some governments are saying, "Hey, you can't just censor us and take our country's intellectual property and just use it. That is not cool. It's not fair. We're pushing back." This is what we need—governments pushing back and governments will push back when citizens push back.

Right now, new legislation is being passed in many EU countries to curb the rights of social media companies. Privacy is being reinforced for humans. In other countries, legislation mandating values and standards to operate social media in their countries is passing their voting bodies and starting to get enforced.

In the US, the relationship between businesses and the government can be problematic regarding freedom and rights for the people. For example, AT&T had a tight relationship with the government for years. It built NSA listening stations in plain sight for the famous AT&T Building in Manhattan which has no windows and is atomic bomb-proof. The NSA has operated a sophisticated intelligence-gathering operation out of the ominous building. The windowless building at 33 Thomas St. in lower Manhattan was busted by Edward Snowden's 2013 data dump which revealed all the shady things our government gets up to. AT&T allows the NSA to listen in on communication in these listening stations—and yes, Americans are being spied on.

It's amazing how Americans can find out that their most well-known telephone company is completely aiding the government to spy on them, and no one blinks an eye. Let me say it louder for the people in the back: America, AT&T is spying on you! How would you feel if someone came into your backyard and was peeking through your window? You would press charges. It's the same thing.

For some reason, Americans have a short memory. And they forgive easily. It's a fast-paced news cycle. People will talk about a problem one day, get all fired up, and move on to the next issue the next week. Why do Americans forgive so easily? Does it have anything to do with money? Or are we trained to have a short memory? Maybe we simply *want* to believe in our institutions? We *want* to believe that the brands we've grown up with, that we know and love, are as faithful to us as we are to them. But surprise—they aren't. And then what do Americans do? They run out and buy their kids a phone to grow up on and love and be faithful to, and they do it in the name of security because they want little Johnny and Suzie "to be safe" and to be able to call them at any time. So, we put the most dangerous device on the planet today in the hands of our children...for their safety. Yes, we have become paranoid since we all got genned up.

# TECH ACTION

1. Do you know when your district's next school board meeting is? Make some calls and make plans to attend the next one. Start being informed and start speaking up.
2. Do you have a list of your local community leaders and how to reach them?

Scan this QR code for additional information
on the topics in this chapter.

markstross.com/book-ch-2

# CHAPTER 3

# The Hacking of Critical Infrastructure

Our free-market society depends on large companies to provide most of our utilities, food, goods, and services. While we're busy learning how to protect our privacy and buying insurance policies for identity theft, we must now think about the companies we depend on and how they should do the same. Otherwise, it can be catastrophic. In this chapter, I'm going to share a few stories that you need to know about because they reveal a typical pattern of corporate negligence, vulnerability, and risk that could have devastating consequences for you, our nation, and our world. If the companies we depend on fail to protect their supply chains from cyber attacks, sabotage, or natural disasters, it could cause massive disruptions, damage, and even loss of life. That's why we need to hold them accountable. We must demand more from the companies that provide essential services and products. Our critical infrastructure companies must do more to provide security for their product supply chain.

The first story is about a cyber attack on the Colonial Pipeline. This is the largest fuel pipeline in the country, which carries about 45% of the East Coast's gasoline, diesel, and jet fuel supply. The Colonial Pipeline cyber attack was one of US history's most significant and disruptive ransomware incidents. The attack was carried out by a criminal group called Darkside, which infiltrated the company's IT systems on May 7, 2021 and demanded a ransom to restore access. In response, Colonial Pipeline proactively shut down its entire pipeline system to contain the threat and prevent further damage. This caused widespread panic among consumers who feared fuel shortages and price hikes. Long lines and hoarding behavior were reported at gas stations across several states, and some stations ran out of fuel completely. The US government declared a state of emergency and took various measures to ease the situation, such as waiving some regulations, issuing temporary waivers for fuel transportation, and providing guidance and resources for states and industry partners. Colonial Pipeline paid the hackers a ransom of about $4.4 million, hoping to restore its operations quickly. However, the decryption tool provided by the hackers was slow and unreliable, forcing the company to rely on its own backups to restart the pipeline. After six days of disruption, Colonial Pipeline announced that it had restarted its entire pipeline system on May 13, 2021, and that fuel delivery had commenced to all markets. The company said the supply chain would take several days to return to normal.

This attack exposed the vulnerability of critical infrastructure to cyber threats and highlighted the need for more robust cybersecurity measures and collaboration across the public and private sectors. The Biden administration launched several initiatives to address the ransomware challenge, such as creating a task force, establishing a collaborative center, issuing security directives, and imposing sanctions on entities linked to the attack.

The second story is about JBS. JBS is in the meatpacking and distribution business and is responsible for 23% of America's meat. JBS is not a brand you probably know, but you would likely know some of its auxiliary brands like Pilgrims, Swift, Great Southern, Aspen Ridge, and approximately 18 other store brands. Well, not long ago, their servers got hacked. Surprise. A sophisticated cyber attack targeted their computers, and some of their meatpacking plants in the United States were closed. This one attack took down 23% of meat processing in America, not only impacting America but also impacting the food supply chain worldwide. This is a big deal! When JBS gets hit, you're looking at global food shortages. Food shortages plus inflation equals disaster for the average household. The hits to our infrastructure create price hikes, which have an inflationary effect. The company ultimately paid $11 million in ransom to prevent further damage, but not before exposing the vulnerability of a big-league American food distributor.

The third story is about SolarWinds, a software company that provides network management tools to many government agencies and private companies. In December 2020, it was revealed that a sophisticated cyber attack by a foreign adversary had compromised the company's software updates, allowing the attackers to access sensitive data and the systems of thousands of customers, including the US Department of Homeland Security, the Treasury, and the Pentagon. The attack was one of the most significant and most damaging cyber espionage campaigns in history, and it exposed the vulnerability of our public and government communication systems.

So, the US got hit in fuel, we got hit in food...what's next? Banks failing of course. All the essential industries, utility, energy, food, and government must invest in better cybersecurity practices and technologies. There should be collaboration with law enforcement and intelligence agencies to adopt a proactive approach to prevent and respond to cyber attacks. These companies must also be transparent and accountable to their customers and the public about their security measures and incidents. If they fail to do so, they risk losing trust and business and ultimately endanger national security and public safety.

When it comes to regulating vulnerability in technology and Big Tech, Congress and the federal government have not picked up the gauntlet, but the states have. States are changing the game with legislation tailored to their politics. Take the TikTok ban in Montana, for example. According to various sources, Montana is a trend-setting state that bans TikTok on government devices. Gov. Greg Gianforte signed the ban into law on May 17, 2023, citing national security concerns over the app's ties to China. Since then, at least 37 other states have followed suit, either partially or fully restricting the use of TikTok on state-owned devices. Some federal lawmakers have also proposed a nationwide ban on the app, which has over a billion downloads worldwide. TikTok has denied sharing user data with the Chinese government or censoring content at its request. This is how states are making a difference, which means it's "power to the people" at a more localized level that is changing the world and will change how it is perceived. Infrastructure must be protected at all costs.

The bottom line: We know that these hacks are effective and hackers are getting behind the firewalls of the companies responsible for our critical infrastructure. JBS profits when we buy their meat, but they are not spending enough on their infrastructure to secure it from harm. This must change.

# TECH ACTION

1. Are you a business owner? Do you have friends who are? What measures are you taking to ensure your infrastructure isn't hacked?
2. Reach out to your state leaders. Ask what is being done to protect critical infrastructure from cyber attacks.

Scan this QR code for additional information
on the topics in this chapter.

markstross.com/book-ch-3

# CHAPTER 4

# The Power of Being Informed

## (and Misinformed)

*Taking our power back as humans begins*
*by taking our privacy back.*

As humans, we gain power by being informed, but we also lose power by being misinformed. In short, we exchange power for information. How? We exchange our power by giving up our privacy. So, to regain the power we've exchanged, we must regain our privacy!

Yes, taking our power back as humans begins by taking our privacy back. To take our privacy back, we must be informed about who has access to our lives and data, which means technology platforms need to be transparent and step up to the plate and protect our privacy. Are there any platforms that are going to hold up against scrutiny? Are there any better choices out there than the social media incumbents? The answer is, yes.

One example would be the messaging app Signal. A gentleman with a crazy name, Moxie Marlinspike, created Signal. Moxie Marlinspike created end-to-end encryption for messaging apps and developed a company called the Open Whisper Systems Company. His technology was introduced to society as "open source," which means he allowed this technology to be used by other companies. Many major messaging apps like WhatsApp and Messenger use this technology today. The Open Whisper System Company donated the source material to create Signal, WhatsApp, and other apps with this encryption.

As a youth, Marlinspike was quite destructive in his way of protesting against Big Tech and government spying. He was a rebel by nature, but in 2015, he decided to make a difference by being a rebel his way. He thought, "Wait a second, I can turn this whole thing upside down if I just create the most intense privacy algorithms out there." He didn't want the government to look at everything we were doing, and many people agreed.

What's nice is that Signal, which is X's biggest rival, has 50 million downloads at the time of this writing. Of course, it's not even close to the number of X subscribers, but it's a start. What's beautiful is that the FBI contacted Signal for information on a person of interest. They said, "Hey, we're conducting an official investigation of an individual, and we want you to give us all your information on them." I am pleased to tell you that all that Signal had was just a name, the date of the origination of the account, the telephone number the account was used on, and when the account was last logged into. That's all they could give the government, period. They had no other information because they kept no other information. For the first time, the government was shocked to discover that a data-driven social media company didn't have much data about its users to monetize. At first, the FBI didn't believe them and thought they were "messing around," and they went back saying, "You must know more about your end-users than you're saying you do." The FBI flexed its legal muscles and tried to force the social media platform to produce the requested information it claimed it did not have. The ACLU agreed to take their case, which is unusual today, to see the ACLU come and protect the free speech of companies leaning outside the political sphere. The ACLU defended Signal, a secure texting app, against an overly broad FBI subpoena that demanded too much information about its users. The ACLU argued that the government had no right to request data like communications metadata, cookie data, and contact lists with just a subpoena. The ACLU also challenged a gag order that prevented Signal from speaking about the request. The ACLU helped Signal protect its users' privacy and civil liberties from government intrusion. That's awesome!

The tides are turning, and the new stuff gives us a separate digital highway to travel. The new highway is proving to be quite efficient at keeping our privacy, thus protecting our power as humans.

For the record, I want to point out something. Some people believe the government ultimately paid for the creation of the Open Whisper architecture. When you look under the hood—and I encourage you to investigate this on your own time—you will discover that Open Whisper Systems funding ultimately came from government grants for projects creating encryption. You must always ask yourself: Where did the money come from? Suppose the money traces back to our government, even if the encryption technology is open source. In that case, we can probably assume that if the government was motivated enough, it could break into it. However, the beautiful thing is that companies are taking technologies like end-to-end encryption very seriously and using them. When these companies are asked to provide information to the government, they decline because they don't have the information. I think this is the healthiest start. But it's only a start because there is also a back door. A back door is a coding reference for another way to enter a secure pro-gram to access its coding. It would be delusional to think there are no back doors in some way or another. Governments pressure Big Tech to give them access; Twitter files brought this into public light. Ultimately, I think those back doors will come at too high a price, even for the government. If you want something to be ultra-secure, it has no back doors. That is tomorrow's issue. Today, the world has started building a counterculture revolution called privacy! I'm super excited!

The other exciting thing is that we are marching to a new place as a culture. We are marching toward having many Internet highways, which will be protected by various groups for various reasons. I think it's a much better situation than what we had five or ten years ago or when the Internet first started. In the beginning, there were no unique highways. This is why the social media app Parler was deplatformed by Amazon. Parler had one platform option and built a whole business on this one highway to the Internet. Amazon disagreed with the content they were posting, so they shut them down. Can you imagine? How ter-rible! Amazon, Apple, and Google took down Parler because they didn't like the content on Parler. During the early 2020s, why did Twitter get to keep North Korea, China, Iran, and other bad actors tweeting and hating on America with state hate speech and threatening to destroy America? Yet Parler's "offenses" were considered deplatform-worthy? The real "problem" was that Parler catered to the "wrong" political party and therefore Amazon deemed its offenses terminal. The truth is that Parler was targeted by a politically-biased set of organizations, who incidentally donated an overwhelming 99 percent of their donations to the opposite political party. Parler was a victim of political censorship. Parler survived after it was taken down but has never truly recovered.

I think shutting down the opposing voice will be much more difficult moving forward because new companies are rising and filling the void. They are saying, "We'll never take you down. You are secure." I hope this continues.

To sum it all up, I'm sure the government can probably break into most software systems, but the more highways we have where they're not allowed to break in, by default, the more difficult it is, though not impossible. After all, when it comes to anything in cybersecurity, I've always said it's not about keeping "them" out forever, just for now. We can never rest on our current security strategies because of the evolving nature of technology.

## The Right of Repair Movement

Now, let's talk about screen protectors. I realize this is such a thrilling subject. And you might be thinking, *Why did we change the subject to screen protectors?* Well, for a good reason. Today, if you use your screen protector wisely, you're protecting your investment. We all know that. Most people don't realize that if you have an "oopsie" and you drop your phone and break your screen, you might as well have gone to a casino and gambled your money. If you try to get a replacement part for that screen and it is not a genuine part from your manufacturer, they have now put something inside new parts called *digital rights protection.* If that part does not match, that phone will not work precisely the way the old part worked. Imagine you have this beautiful phone, you broke your screen, then replaced it with a generic screen, and now you have diminished capabilities on this device because the part was generic. Suddenly, that screen protector becomes quite significant. This is the first point I want to make about screen protectors. The second point is that if you break your device and you are not using a screen protector, you now get into a more significant issue called *the right of repair*. This is a big topic, but the truth today is manufacturers would prefer you not be able to repair your device anywhere but at their shops. That's not good for the consumer because, after three years, most manufacturers won't repair your device anyway. If you have an older device, that device is no longer repairable. You cannot bring that device back to the same level that it was before requiring the repair because if you don't have the right part, it won't work the same. That's a big issue, and I am for the right of repair. I don't know about you, but I want to own the device I paid for. I like to choose who can repair it and when. I don't think anyone should tell me I can't repair my property. More importantly, if you're spending over $1,000 on a phone, you should be able to have a reasonable expectation that you can fix it if it breaks. For example, imagine you're on holiday and don't have an Apple repair shop where you are, and you must go to a generic shop. It would be nice to be able to put a screen on your phone if you need to. Unfortunately, today, that's not the "case." No pun intended.

Apple is starting to head down the right path, but we have a long way to go, and while all these companies are screaming "go green," no one is green if they don't recycle and fix gear instead of making it disposable.

The right-to-repair movement is centered around the idea that individuals should have the right to repair or modify their products. This includes everything from smartphones and laptops to tractors and other large machinery. At the heart of this movement is the belief that when individuals purchase a product, they should have full ownership and control over it. The right-to-repair movement advocates for laws and policies that protect consumers. This includes requiring manufacturers to provide access to repair manuals and diagnostic tools and allowing third-party repair providers to access parts and software. This promotes a sense of empowerment and control rather than the feeling of leasing a product you cannot fully own or change.

Now, I can't speak about every brand. I can only speak about some of the flagships. Every brand of smartphone has different policies regarding repairability. Your Google phone might "play nice" about one form of repair, but many brands might not do the same repair. You don't want to find out what the repair policy is after the fact because, most often, the policy will not cover whatever accident you just had. This is my warning to everyone: put those phone cases on. Put your screen protector on. Don't worry about the fashion statement and protect your device because manufacturers are going to give you about three years of repairability on average. Then, you're on your own in the Wild West where likely your device is not repairable.

## Everyone is the Solution

In 1775, Thomas Paine wrote an essay about the balance of power and the scale of peace. When Thomas Paine wrote his essay, "The Force of Defensive War," he talked about giving arms to the people and why that was important because if you had a government that became too powerful, the government would take advantage of the weak. With an imbalance of power, the strong prey on the unarmed weak. While Thomas Paine was talking about our right to defend ourselves from the government with weapons, in this technological age we need to protect ourselves digitally from Big Tech and big government.

What blows me away about this essay is realizing that one of our founding fathers could predict what we would be dealing with today, even though he had no idea of the technologies to come.

Should governments snoop through an app where there are both criminals and innocent people and read all their communications? Next, is a story that brings to the forefront a fantastic shift in technology usage in law enforcement and their snooping capabilities.

Although we have incredible encryption technology, criminals have been able to hide behind encrypted firewalls, which act like a digital privacy fence, usually found on the dark web. For a long time, criminals have evaded justice because these encrypted firewalls are so secure. However, a profound shift in the balance of power occurred around the world when law enforcement agencies from multiple countries came together and devised a plan to bring down those encryption barriers— the first global sting operation in history!

It started with law enforcement getting ahead of criminals by breaking into their secure chat rooms. Based on those hacks, they developed an incredible plan called Anam. Anam was developed by a "white hacker," a gentleman coder who used to be a dark hacker up to no good, but he had a complete turn around and is now helping the police and the FBI. The European FBI and the Australian police force came together on this operation and decided to do something novel. They were going to figure out how to do a digital sting operation. Law enforcement worked with hackers and said, "Wouldn't it be cool if we could have control of the communication platform that criminals are using?" They all agreed and began to get to work. They created encrypted phones with back doors to give them access and sold them to criminals on the black market. Undercover agents convinced prospective buyers that these phones were secure, and once the criminals bought in, the police had a foot inside these criminal organizations. That was the first step. The criminals sold the phones throughout their network, and things got interesting. Law enforcement could completely infiltrate these criminal networks by hacking into their technology. No one knows the number of phones sold, but it's estimated around 800 phones went out.

Interestingly, these phones all had to have subscription plans to operate, so the criminals were paying for the phones, which means they were paying for the FBI to track their criminal organization. Then what happened? With these phones, for the first time, the FBI was monitoring in real time a shipment of cocaine coming into the US. They were also reading messages about criminal activity such as murder plots. As a result, they seized $45 million in cash and apprehended over 800 criminals globally. In Australia, 20 million messages were reviewed leading to the prevention of over 20 murder plots in Australia alone, and six other murder plots worldwide. Around 4,000 police officers were involved in this global operation.

I find it even more impressive that this operation happened worldwide in Asia, Australia, Europe, South America, and the Middle East. These law enforcement organizations were able to operate together. For three years, these phones were sold on the black market, and the secret was held by many law enforcement personnel. That's wild! Criminals didn't figure out that those phones were exploiting them, which is phenomenal because, if you think about it, these guys were making a billion dollars in Australia alone. They were brought down because they could not secure their communication.

They bought ultra-secure phones on the black market, and they never did a background check on the phone itself. *They just assumed it was an encrypted phone.*

Of course, we are happy to see criminals brought to justice, but the techniques used here in this case are also the techniques that can be used on the public if encryption is taken away from the crowd. It's our last defense of privacy against the government currently.

My question again, and I don't have a perfect answer: Should the government be able to snoop on us and our communication? Is that okay? Is a search warrant okay if its coverage is too broad—like a search warrant to cover all the communication on a chat app? Is it alright to start snooping on innocent people more broadly to find a criminal? Something we all need to monitor and decide for ourselves.

## UFOs: Who Controls the Narrative?

We can't possibly talk about being informed without talking about UFOs. So let's start with *The X-Files*, a television show that aired from 1993 to 2002. I must salute them because they were among the first shows to talk consistently about the giant "800-pound gorilla" that no one else seemed to be talking about—UFOs. Then came *Ancient Aliens*, *In Search of Aliens*, *Project Blue Book*, *UFOs*, *The Lost Evidence*, *Alien Highway*, *UFO Hunters*, and *Unsealed Alien Files*. Then, finally, *UFOs Declassified*. Now with all these shows—I remember watching most of these shows—I always thought, *When will we finally have someone declare this evidence is objective? There's all this photographic evidence, and when will it be certified?* Well, the truth is coming out now. The problem is that we have been so misinformed that most people wouldn't know the truth if a green Martian were staring them right in the face!

I have a little story for you. In the 90s, I was working on the production for a television show in LA called *Mysteries from Beyond the Other
Dominion*. In this series, the main character was Dr. Ruehl. Dr. Ruehl was
a scientist and a ufologist. It was one of the first shows on the Science
Fiction Channel, and it was ours. It was a very low-budget show, but
in gathering information and footage to use in the series, I got to see
evidence of UFO sightings and other strange events that did not have
explanations. For example, we had footage of things in space making
90-degree turns. Now, NASA explains that away by saying they were ice
crystals making 90-degree turns as seen from the space shuttle. But I
wasn't satisfied with that answer. I wanted to analyze the footage further.
I had made friends with the computer scientists at Caltech and took the
footage to them. What is fascinating is that even my friends, the scientists, were not convinced that ice crystals could make 90-degree turns
in space. I mean, how was this ice mysteriously making radical U-turns
*in space*?

Most everyone seems to have a theory about UFOs. Whether we
all agree or not, it doesn't matter. The point is that something we can't
explain is going on. The military has finally tracked these weird things,
traveling 13,000 miles per hour. In 2021, the world was stunned when
*60 Minutes* revealed that UFOs had been spotted by US Navy pilots
and confirmed by Pentagon officials. The report featured interviews with
eyewitnesses, experts, and former government officials who shared their
views on the mysterious phenomena. The *60 Minutes* report raised many
questions about the nature and origin of these objects and whether they
threaten national security or human civilization.

The actual physics of what these crafts can do is fantastic! They can
travel 13,000 miles an hour, pull off 600 Gs, evade radar, and fly through

any element. Their speed and velocity don't seem to change if they're dipping into the water, going underneath it, in the air, or in any form of air turbulence. None of our present technologies can do that!

In 2021, the military came out with pictures and video of their UAPs—Unidentified Aerial Phenomena (the new trendy name for UFO). We know today that our current understanding of physics can't explain these phenomena. Equally interesting, some scientists claim that the instruments we use to observe these unidentified crafts create unusual imagery, and we are seeing not the movement patterns or characteristics of the UFOs but misdiagnosed signals. I have seen the footage personally, and as a video and technology expert, that does not appear to be the case. I think these are pictures of natural, unexplained phenomena, and I am okay with not knowing everything about our world or what's in it. We still must go out and explore. As said by Captain Kirk, "We must boldly go where no man has gone before."

Now, let's return to the facts and forget all the controversy. The facts are straightforward. The military has tracked items. They have no idea where these items come from or how they operate. Is that a bad thing? No. Is it wrong that we don't understand the propulsion of these things? No. Is it a bad thing that we don't understand any of it? No. It just means we have something yet to learn.

People tend to view UFOs as a black-or-white issue. You're either in or you're out. I'm entirely and very comfortably right in the middle because, from my perspective, these UFOs could be another form of physics we know nothing about. They could be based on a completely different life form outside our understanding of physics and propulsion. Right now, these items seem to "play" with our military jets or anything with radar and lots of electronics attached to them. What happens if they're like fireflies, just attracted to electromagnetic energy and playing with us? They've done nothing hostile. They seem to be playing. If we have planes in the sky and the plane starts to dogfight, they seem to dogfight with us but then go away; they have done nothing to harm us. So far.

Another important thing to consider is that we know the military has been dealing with these unusual aircraft for many years. I expect that the military has taken these sightings seriously. I believe that if these were UFO military craft threatening our restricted airspace or hostile in any way, per the *60 Minutes* interview, I believe our military would be on an all-out mission to discover what these things are with full-blown transparency.

The most crucial observation is that, before this revelation, it might have been thought that the whole planet would freak out about this news. But the government has revealed everything, and they didn't put a nice little bow on top of it either. They said, "We don't have all the answers."

These transactions have one thing in common—they've been benign. Therefore, from my perspective, I look at UFOs as completely real and completely unidentified, and I am entirely okay with it. I have many theories of what they could be, but again, those are theories. It doesn't matter that all of us have our theories. What's most important is that we accept things even when we don't understand them. The premise of *The X-Files* was about accepting the possibility that there is more out there than we know. We have a ton to learn.

I'd like to see society becoming comfortable with the fact that UFOs exist because it means that our science is still young, and we have plenty of places to go. Just think, not long ago we would watch all those television shows I just named, waiting for the evidence. "Please give us some evidence!" Now, the military has said, "Yes, we're tracking these things, 13,000 miles per hour, and we've got nothing like it. We don't even know what they are." And for the most part, the public is mute. No one is talking about it. Why is that? Is it because we've been misinformed for so long that we don't know how to react when they present information contrary to our beliefs? Friend, this is known as cognitive dissonance—the gap between our thoughts and reality. We're on the verge of probably one of the most significant scientific discoveries in humanity, and the public has essentially looked the other way. We're not even talking about it! Well, I am, at least. Most people have just said, "Oh, ok." And then buried their faces back into their devices.

I hope today I can inspire you to start thinking about this issue and take an interest in it. If you are a young scientist reading this, it is an excellent subject to study. I would love to see us bring UFOs out of the closet. Let's go from thinking, "You can't talk about that because it's just weird," to "We don't talk about UFOs enough." Just understanding their flying physics could change our world forever. We need to go on an adventure to figure this out. It is pure hubris for us to assume we know everything. It's hubris for us to assume "It's aliens" or "No, it's nothing" or "It's something." The point is, we don't know. We must be okay with uncertainty, or we will cling to specific ideas even when they aren't the truth.

# TECH ACTION

1. Look at all the apps you have on your phone. What are the privacy settings? Do you know who is selling your information and spying on you?
2. Do you have a screen protector and a good case on your phone? Do you know what rights you must have to repair your device should you need to?
3. Find a friend or loved one and discuss the discovery of aliens. Do some of your research. Don't be afraid to be informed or change your beliefs when new information is presented.

Scan this QR code for additional information
on the topics in this chapter.

markstross.com/book-ch-4

# CHAPTER 5

# The Unseen Cost of Technology

*Not only can technology be used against us,*
*it can turn against us.*

Can you even remember life before cell phones? I remember using an atlas and a payphone on a long drive. I remember waiting to use the phone at home because it was a party line. We've come a long way with our tech which has made life so much simpler. We as a culture have become very reliant on technology, it's true. However, too much of a good thing can be a real problem. Here's why: Technology can do almost everything imaginable without a moral compass. We treat our phones and digital technologies as smart, but are they? They will give us anything we ask for, but there is one aspect of this transaction that we never think about. When you look up material or use an app, you are at the mercy of its creators and their worldview. We seem to naturally assume that our morality is transferred into everything we do, including what we do digitally. But in practice, it's the other way around. The true intent of an app is designed by its creator. Our morality is being side-lined by institutional group morality, and this has occurred as we have become more dependent upon apps to run our lives. Many people tout all the good that technologies have done for humanity, and there is a lot of good. But we cannot overlook the other side of the coin—where individual morality, ethics, and happiness are being eroded. Since the smartphone's introduction in 2010, people's general happiness has been in freefall around the world, while depression and anxiety are on the rise.[2] Why? Is it because we are being corralled into smaller and smaller echo chambers so Big Tech and app developers can impose their values on us? I would say, unequivocally, "yes."

Collectively, we are allowing our humanity to be replaced by rigid rules of engagement designed by the creators of our technologies. Imagine how many apps you use every day. Think about the fact that you have agreed to the terms and conditions set forth by every single one of them. What have you agreed to? Did you read those terms and conditions? And does it matter if you did? There's no place to share your opinion and redline the contract. If you want to use the technology, you must opt in. In this environment, our individuality is being opted out. Never has society been so locked into new rules that we must follow to make everyday lives work.

And speaking of eroding morality, the issue is far more significant than you would think. When Facebook was created in the US in 2004, it took China and other countries about a year to duplicate it. Why is this a problem? Because other countries don't have America's best interest in mind. Let me tell you a personal story...

There was a time in my career where I was traveling worldwide,

---

2       https://www.cancer.gov/about-cancer/causes-prevention/risk/radiation/cell-phones-fact-sheet
https://www.health.harvard.edu/blog/is-your-cell-phone-dangerous-to-your-health-2020012118697
https://www.vox.com/2018/7/16/17067214/cellphone-cancer-5g-evidence-studies
https://www.ncbi.nlm.nih.gov/pmc/articles/PMC5504984/
https://ntp.niehs.nih.gov/whatwestudy/topics/cellphones/index.html

meeting with the ultra-wealthy about my video board technology. One of my meetings was in Dubai, United Arab Emirates. By this time, I was already working with a holdings company to bring video boards to the region, and I was going to control the video boards with the software I invented. At least, that is what I thought was going to happen. While I was there, I was invited to a mysterious meeting. I was picked up from my hotel by a Rolls Royce and chauffeured to a dock, where I was then escorted by a security detail to the main stateroom aboard a huge yacht. At this point, no one spoke to me or told me whom I was meeting with. The stateroom was filled with gold items, fabrics of deep greens, red velvet, and silk with a distinct Gucci feel. It was both opulent and functional, and I was ushered to a chair near a small round coffee table. A gentleman came out of a side room and sat in a chair facing mine. He didn't say a word but snapped his fingers, and immediately, two cups were placed before us like magic! Then, the gentleman poured something like black tar into each cup. I was asked if I needed sugar; I declined. The man across from me smiled and said, "You like your coffee as I like it." I smiled and held up my cup and drank. Boy, did it taste bitter and stout—like caffeine-infused mud. Motor oil might have tasted better. After I took my sip and smiled—even though my taste buds were under assault—I put down my cup and said, "Thank you for your hospitality; how can I be of service to you today?" The man looked at me and said, "Mark, I do not need your services. But I have decided that you should not be hurt." He said it with a direct, deadpan stare. He was not joking. "Mark, do not give us the technology to run your video boards," he said. I sat there confused and slightly terrified. He continued, "Our original plan was to learn from Westerners and then just 'dismiss' them. However, you have impressed my team and me with how you approach them and care about their outcomes. If you can change their hearts through actions, I will help you. I decline to do business with you because it would hurt you."

That was the end of the meeting. I finished my black sludge and thanked the man, still today not knowing who he was. But I was grateful he decided not to "dismiss" me. Tickets for my trip home, first class, and final project check were at the hotel. My stay in Dubai was over.

A lesson learned: good works sometimes give you gifts that are not apparent now but will reveal themselves in time. But as it relates to the topic of this chapter, just know that other countries do not care about America or its advancement.

## Duped Because it's Duplicable

Original technology vs. duplicated technology is very different. The original tech usually has the heart and soul of the creator in it, with a sense of morality about how the tech is consumed. This is why the creator should maintain ownership of that intellectual property. When computer code is copied, it is stolen. The people who are copying it are doing so to make money, and the inherent morality of the product is removed. This is another massive problem with technology—it's duplicable and becoming more accessible and easier to duplicate.

If you think about everything from creating fake handbags and sneakers to professionally cloned software, it should all be criminal. And it is, sort of. Unfortunately, we in America today face a foe called China. China has a different political system, and they don't have any antitrust laws like we have in America to protect consumers. China doesn't have any provisions in place to stop an organization from super-amassing assets and power. If the corporation shares with the Chinese govern-ment, everything's copacetic.

In China, entities can grow alarmingly big and expand into all types of industries without restraint. For example, a social media app can take over banking, transportation, delivery, and entertainment, in addition to its original social media functions. So, the app can now do as many things as the user needs or wants for their daily life, and it gets a unique perspective about the user. The app takes the collected information and creates a complete three-dimensional picture of your identity, what you like, where you go, what you do in your social time, and, more impor-tantly, where you have been and will go. Yes, the all-in-one app means that your life has a complete digital footprint that any government can easily use to figure out what you have been doing and thinking and what you will do next. Does that sit well with you as you read this?

Now, if you have a totalitarian government like China with an Orwellian high-tech surveillance system capable of monitoring every keyboard stroke, is that an invasion of your privacy? Hint: YES! And what if those governments decide to take your data and use it against you because there are no rules or regulations to protect you? You would undoubtedly find yourself in a very bleak spot, to put it mildly. Now, you suddenly find out what you have exchanged for your smartphone—the actual cost for the phone in your pocket is your freedom. Why is it that people treat their phones like a lover they can never let out of their sight day or night? Or like a trusted companion with whom they consult about every single decision? The truth is that the one who stood by you and gave you everything you needed, the one you depended on and talked to all day and night, your precious, is not a friend. It's more like a secret agent, and it's been spying on you all along. Maybe they should call it the "outsmartphone" because it's outsmarting all of us!

The most crucial point is the erosion of trust. Americans don't trust each other, let alone trust technology, and more importantly, we no longer have moral context for technology. That, to me, is super scary. The more technologies are copied and duplicated, the more people feel duped and truly, it's hurting all of us. Let me give you some examples of how this is happening.

China built an airline jet called the C919. Many experts claim that the plane was constructed via espionage—stolen technology from around the world rather than good old-fashioned R&D. (That's research and development, not rip-off and deception!)

If that's not enough to make you cringe, not only is China stealing technology, it's using it to control its people. (Sounds like something I may have warned about earlier in this book.) China is now using social credits in social media to "honor" citizens who are "good" citizens or compliant. If you are not obedient, your social credit score drops, and if your social credit score drops below a certain point, you can no longer rent cars, take trains, or function in society—just like it was depicted in an episode of the TV show *Black Mirror*. Black Mirror is a UK and US non-hosted anthology series that unveils how modern technologies can backfire and be used against their makers. The "Black Mirror" is technology. Every episode is set in a slightly different reality, with other characters combating different types of technologies, raising questions of morality and ethics. The series ran from 2011 until 2019. In one episode, an individual could not participate in her society because her social credit score was too low. This is no longer hypothetical. During COVID, China moved its whole social structure to the social credit system. They used COVID as an excuse to ratchet up the Orwellian surveillance technology. When science fiction becomes fact, it's eye-opening.

That's not to say that America isn't doing the same thing regarding pirating technology. I mean, no one knows who steals what. It's called hacking for a reason. However, I do not believe America has stolen much from China because we've basically created everything they've needed and used. Here's a kicker—if you want to do business in China, you must allow a Chinese company to share your intellectual property. It is hard to believe, but every tech company, including Apple, has agreed to those terms and ends up sharing their intellectual property with China. Therefore, China gets a huge boost! They haven't had to invent very much—we did the inventing; they just cloned it.

As I explained earlier, this is the fundamental problem with cloning something. You, in essence, do not intellectually comprehend or under-stand how the technology works. Unfortunately, China feels that it is perfectly acceptable to build a plane and steal as much tech as possible because, in their psyche, they believe that Western societies have taken advantage of China for hundreds of years, so it is payback time. They believe intrinsically that they have a right to this technology because they feel that the US oppressed them originally. China is also poking proac-tively at Japan and other countries by infringing on their sovereignty. China is exerting its influence that it believes psychologically it has a right to, which is a mindset that most Americans don't fully comprehend. China believes they were an oppressed country, specifically by Japan and other countries during World War II, and were walked all over. They want to come back as a superpower and believe they're owed this technology from Western companies. I've worked with captains of industry in China and have heard firsthand this perspective. Seven years ago, I was told by the president of a Chinese LED company that basically America was over, and China was going to bury us. I remember smiling and replying, "The only problem is that you only know how to copy. We know how to create. We still have heart. You might try to bury us, but our hearts will prevail." There's good news: China is no longer the world's factory. People are pulling out and bringing industry back home.

I want Americans to know that if anyone thinks for one second that China is passive about this mission to take us over, it isn't. This is an indoctrinated concept that they will acquire all their tech in any way possible, and it's all on the table. Americans don't operate that way, and that's the genuine existential threat.

So why would American technology companies agree to give up their intellectual property? Because China has over 4.5 times the population of the United States. Which means a company has a lot of potential sales in China. Yes—it boils down to money. If for every billion dollars you can make in the US, you can make 4x more in China, that's a proposition many corporations can't pass up. It's hard to believe that a company would sweat over a patent, create a product that they want to manufacture, and then they would hand over their blueprints to another company. Just because they want the largest market share, they're willing to blow up the actual jewel of their creation. Honestly, I don't understand it. By the way, up to 2021, this was the way American companies were operating. I am pleased to announce that this has changed, moving into 2023. China is declining as the world's number one manufacturer, and other countries are slowly ramping up to replace them. Guess what? A country on our border is slated to be the "new China" for manufacturing—Mexico.

Because of COVID-19, in China today, we have WeChat and other technologies that are now entirely plugged into the populace. These apps have become mega apps. They do everything from banking to new entertainment, as I mentioned before. Imagine you have an app that tracks your location, finances, what you eat, what restaurants you've visited, tracking new entertainment, and everything. Well, that's what's going on. Everybody now has a spy in their pocket tracking everything. And if you don't think the CCP is using data against people, think again. They have already used the social standing system to embarrass people to fall into line with COVID mandates. They have used social standings to control whole cities and lockdown cities so precisely that it would put Hitler's armies to shame.

Now, China believes that they have moved society forward with this technology and are actively implementing their Silk Road initiative. The Belt and Road Initiative (BRI), also known as the One Belt One Road (OBOR), is a global infrastructure development strategy adopted by the Chinese government in 2013 to invest in more than 150 countries and international organizations. It is intended to link China with more than 100 countries through railroad, shipping, and energy projects. This is their way of trading out to the world. They want to trade technologies with the world and for all the countries in their sphere to use similar technologies. Ultimately, this is about building the power of the Chinese government over the world, one country at a time.

One significant caveat in all this is that we must consider that China does not honor their agreements. Okay, I can hear someone saying, "Well, Mark, there, you just said something racist or bigoted or against that country or xenophobic." No, I have not. The Chinese people are great. I have worked with the Chinese people. The difference is that the Chinese Communist Party and the ruling class of China, mostly CCP members, don't think the way the rest of the Chinese people think. If you take away religion from a people, and you take away ethics, and you inspire people through communism, those people will have a lower moral threshold. In

China, ethics have a completely different weight from Western ethics. Please read this carefully because this is super important, and most Americans don't consider that the Chinese legal system, their government, has created a standard of ethics acceptable in China. Those same ethics would not pass muster in America, but in China, they're okay because of their political party. When dealing with the Chinese government or a corporation with ties to the government, you're dealing with a completely different system with different ethics and morality. When we say in America, "We would never do that to you," we wouldn't. But in China, they might. Lying to get what you want is considered honorable over there.

Now, in America, we have one incredible institution that's protecting us right now: our Constitution. We also have antitrust laws. In America, we've decided that no company should become monolithic, and monopolies are controlled through antitrust. Because of antitrust laws, the saving grace is that Meta can't expand to every industry because it would be considered too large, and we tend to break up companies that get too powerful. This protects consumers.

However, what's happening now is that these would-be monopolies are going to American legislators saying, "That's not fair. In China, they're getting larger than us and more powerful because they don't have this regulation." They're trying to influence our government, saying, "Give us the ability to go into banking, go into entertainment, go into all of these sectors, and we'll only create a better life for human beings." And that may be true. These super apps help people's lives...until the government uses them to shut down people's lives. The dilemma for America is ease of use. Are these super apps going to be permissible, or will we hold the limits on how much power a single corporation is allowed to have?

Every citizen needs to think long and hard about what may come if American antitrust laws are changed. How much power should an organization have over you? If the government seizes the assets of that organization, would they have unprecedented information on which to exert authority? This is where we are today. We're at a decision point. Every American citizen has the power to make a difference. I've said this often: We're beyond asking our government to help. The government is not coming to help in this situation. US citizens must tell the government, "We've had enough!" We want to see limits on power and ensure that freedoms granted in the Constitution are upheld and corporations stop having unbridled access to us as human beings. This starts at the state level.

I want to remind you that Meta and Google are using an advertising model where you're just a product. You do not want to allow a company that treats you like a product to end up owning you. They will not have your best interest or any of your interests in mind. In China, these super apps give their government complete control. Currently, we only have our Constitution and antitrust laws protecting us from the same fate. America, if we don't act, and we continue this trajectory of focusing on "what's convenient for me" while nobody is paying attention to the consequences, we're going to get ourselves into big trouble. Before we know it, we will voluntarily give our freedoms away. China has a good long-term plan. It seems like America has reasonable short-term goals. It's time to change that and start playing the long game, America.

# The Singularity

I started this chapter with evidence about the level of influence technology has over us and how it can be used against us. But not only can it be *used* against us, it can *turn* against us. While this sounds like it's straight out of a sci-fi movie, I assure you it isn't. So, I will continue with this chapter's theme and address the issue of technology taking over the planet. Now that I have your attention, you should know there is an event that's going to happen, and many people in the world of technology are a little bit apprehensive about it, and for good reason. What is this event? It's called the singularity.

The singularity is the moment in time when AI becomes intellectually equivalent to humans. We are already seeing the exponential rise of software-based intelligence, which will soon surpass any human or group of human intelligence. Then computers can think by themselves. Computers will learn to program themselves without the need for humans. At that point, a question needs to be posed. Who has programmed the AI? What ethics and morality will the new intelligence have? This is very important. Side note: Some call this a new life form, but I would not go that far. I define a life form as organic, with biology and life and death in its equation.

Why is it called the singularity? In math, we use "singularity" to define a non-solvable equation, like infinity and zero or one and zero. What's the space between one and zero? Does it even exist? Things that we can't solve we call "the singularity." As it relates to AI, the singularity is this: if you have exponential computer growth, which equates to computer intelligence, and computer intelligence has no limit, and everything keeps accelerating, that means that right now, with our current understanding of the world and AI, we are in the Stone Age compared to where we will be just 30 years from now. There is no way for us even to fathom what it will be like in 2050. Think about this: In 2000, you didn't have a smartphone. You didn't have the smart television. If you think about it, the Apple and Android revolutions are only about 15 years old. In 15 years, the world has completely changed the way it interacts and connects as human beings. If you take where we are today and extend it, you can understand why a paradigm shift is on the horizon that is so complex they call it a singularity. It is a moment when we can't predict the outcome or consequence.

If you want more information on this phenomenon, I encourage you to go online and look up videos on the singularity. Ray Kurzweil's channel is a great place to start. Some people love him. Some people hate him. But he has good information so you can draw your conclusions.

## The Phasing Out of Humans

You know, it's interesting to think about computers becoming human. As a kid, I watched the movie "Modern Times" with Charlie Chaplin. This movie occurs in a vast factory and depicts a bleak, modern world. Charlie Chaplin is a worker trying to please a demanding boss and not get swallowed up by the machines he must work on. What I love is Chaplin's movie from 1936 is relevant today. It suggested that our machines will be running us one day, not the other way round. That is what's happening today—machines are starting to run our lives.

Right now, we are at the beginning of the robotic age. We don't see it, but robots have already taken over manufacturing. A Tesla factory has more robot labor in it than it has humans. Do you think it will stop there?

Eventually, robots will move into every significant sector of society, and when we see it happening, it's already too late. Already, those human jobs have dissipated forever. Instead of eight grocery clerk jobs, there is one, monitoring the self-checkout stations. If you work at Amazon, your workday is broken down into 15-minute segments where AI manages your time. You are so controlled that you are essentially becoming a robot to Amazon. But then you know you will be replaced by a machine as soon as the business can do it because your job is entirely program-mable. Amazon does not need you; you're just a cog in the machine's wheel. At the turn of the century, that's what you wanted. Working hard in a factory was considered a stable job. Today, we are being kicked out of the factory. You are being replaced.

There is such a disregard for human beings in this situation. The capi-talists and the manufacturers don't care, and the government is so obtuse that most government agencies don't know the magnitude of the problem this is creating. "Rome is on fire," and jobs are going away forever. This has nothing to do with China or Russia. You are in for one trippy outcome if companies don't need human workers. Do you have to provide income for those people who are nonessential workers? Either that or they are out on the streets. And then what happens? Mad Max, anyone? An environ-ment that feels like a post-apocalyptic America? There's an unseen price for technology.

# TECH ACTION

1. Do you have a backup plan if AI replaces your job?

2. When you purchase something, do you look at where the product was made?

Scan this QR code for additional information
on the topics in this chapter.

markstross.com/book-ch-5

CHAPTER 6

# Cyber Warfare
# in Your Kitchen

*You can't talk about national security*
*without talking about the Internet.*

People are unpredictable. That's no surprise. With the failure of Meta's virtual world, we have discovered that people are unwilling to spend unlimited time in a low-resolution digital reality. And in case you've been living under a rock or on a remote island in the South Pacific (which sounds ideal to many disenfranchised folks), Facebook, which became Meta, tried to create a digital world called Meta, spending billions of dollars, and then shelved the whole project. It was an interesting combo of digital representation of stuff we already have in the real world and digitally created stuff that we don't have anywhere else! I'll talk more about that in the next chapter. The problem isn't so much that the real world is being recreated in a digital world; the problem is when activity in the digital world branches out to the real world. In the digital world, there are no consequences and, therefore, no restraints. Ethics are a moot point. It's a free-for-all online. Offline, most people generally don't go around worrying that bad stuff will happen to them. They never think it will be their street, house, room, computer, device, or mind—or worse, their child. On my street, in the real world, a teenager was killed over money and drugs. I lived in an upscale, gated community in a quiet suburb of Houston at the time. The teen's friends all came from middle-class families, and they all got addicted to making money as fast as possible through drugs, and it banked a dead teenager. *Were there signs that this teenager was in trouble?* Yes, if you examined his social media posts, which people did after the fact. There were countless images of him fanning out hundred-dollar bills to indicate stature. How in the world was a teen amassing this type of wealth? That's the question adults should have been asking. Pretty much, only one way. What is sad is that any psych professional would have seen the signs of peril looking at his feed, but the problem is the adults in his life weren't even looking. Only other teenagers, who were a part of the addiction cycle of social media acceptance and the lure of making money at any cost, saw his posts. We won't even get into the role the social media companies play in all of this. Let me explain: social media companies can detect information about COVID and January 6th protest chatter instantly, but on a kid's account, they do not track predators and gangsters' behavior and conversations clearly about drugs. This is not normal behavior. Parents: There is a battle going on right now for the mind of your child or teen. It's happening right under your noses. It's happening in their bedroom. It's happening in your kitchen. Yes, the battle is personal, and it's on your turf. There are all kinds of threats to kids using technology. Aside from destructive temptations and habits, natural online predators are lurking en masse. Please pay attention and get involved in your kid's online experience.

Back in the metaverse, cyberstalking occurs within seconds of a fresh identity hitting the web for the first time. There's a story on

YouTube where a cybersecurity company and the FBI made a fake ID of an 11-year-old. Within 2 minutes of that ID being let loose on social media, predators were already requesting in-person meetings and asking her to take off her clothes and send pictures! Let that sink in. The lousy stuff only takes 2 minutes to begin when your child gets online. The family unit is in a battle, and the war has been ongoing since 2010 when the smarter-than-the-parents smartphones became available to kids and teenagers. Guess who is losing the fight? Parents, families, and our society. It's time to change that.

## Helping the Next Generation

Millennials and Gen Zers are the first generations to have grown up with widespread access to digital technology and the Internet. While this has many benefits, there are growing concerns about the potential negative impacts on these two generations' mental health and well-being. To understand the full extent of the damage that has been done, ongoing philosophical and mental research is needed. This research can help shed light on how digital technology impacts cognitive processes, emotional development, and social interactions.

For example, research has identified several potential negative impacts of digital technology on mental health, including increased anxiety and depression, decreased attention span, and a decline in face-to-face social interaction. However, much is still not fully understood about these issues and the complex interactions between technology and mental health. Much is unknown as it relates to the impact of social media on our sense of self, the ethics of data collection and surveillance, and the role of technology in shaping our values and beliefs. By better understanding these issues, we can work to develop strategies and interventions to promote healthier, more sustainable interactions with technology and protect the well-being of future generations.

Our kids know more about the world today without leaving their bedrooms than their parents ever did. The idea that a parent is more intelligent than their kids may be technically accurate, but in the mind of the child who can look up anything that has been said and find an opposing view in seconds, it takes away respect for the parent. That is a shame. But we can reverse this trend by saying "no" more often. I said it before and I'll say it again. Parents: say "no" to your video games. I realize the armies of programmers have figured out how to make play addictive to all of us. You included. So, it's as much about you saying "no" to yourself as it is to your children. Please—say "no."

A couple named John and Jane (names are changed) had a young son named Michael. They were loving parents who always wanted the best for him. However, they struggled to set boundaries and say "no" to Michael. When Michael was growing up, his parents never denied him anything he wanted. They showered him with gifts and gave him unlimited Internet access without supervision. As he got older, Michael became curious about adult content on the Internet, and despite the minimum age restrictions, he could access pornographic material. His parents said nothing about it, thinking it was a phase he would grow out of. As time passed, Michael's addiction to porn became more severe, and he started skipping classes to watch porn instead. Despite failing grades, his parents didn't intervene. They were afraid of upsetting him or damaging their relationship. Eventually, Michael's addiction to porn consumed his life, and he began to spiral out of control. One day, Michael's school counselor called his parents to discuss his failing grades and increasingly erratic behavior. Then, John and Jane realized they had failed to set boundaries for their son and allowed his addiction to control his life. Michael was sent to therapy, but he was unhappy and unresponsive. He couldn't find a job because of his addiction, and his self-esteem was at an all-time low.

John and Jane regretted not intervening earlier and were heartbroken to see their son suffer. Finally, after months of therapy, Michael started to make progress. With his therapist's help, he overcame his addiction and got back on track with his life. His parents had learned their lesson and started to set boundaries and communicate better with their son. In the end, Michael was able to turn his life around, but it was a hard lesson for his parents to learn. They realized that sometimes saying "no" is the best thing they could do for their child's well-being and theirs.

Parents, make the hard choice now. Be proactive now. Call your local officials directly. Stand up for your digital rights now. Your loved ones are counting on you. Humanity is counting on you.

## Veterans for Child Rescue

I am proud to be part of Veterans for Child Rescue (V4CR), a 501(c)(3) nonprofit organization founded in April 2017. Their team is comprised of military professionals, former and current law enforcement, senior intelligence community veterans, child abuse and trafficking survivors, and people who are willing to do whatever it takes to end child trafficking in the United States (Vets4ChildRescue.org). They conducted similar sting operations with a female adult posing as a minor. She also happens to be a friend. With the help of AI, they made her photos look like she is just 12-14 years old. When she posted the images on different social media platforms, guess how fast it took for predators to send her mes-sages with the most inappropriate, harmful solicitations, images and videos you can imagine? Yes...unfortunately it took only seconds! The good news is this organization is working with local and federal law enforcement to get child predators off the streets and put them behind bars. If the predator incriminates themselves, law enforcement will issue warrants and make the arrests with a 100% conviction rate. This is one of the few times I can see progress against an ocean of harmful digi-tal perversions.

Parents, if you are alarmed, please take the suggestions presented in this book to heart and act on them.

Here are some steps you can take to protect your children online:

- **Educate**: Teach children about the dangers of the Internet and the importance of being careful about whom they communicate with online. Encourage them to talk to you or another trusted adult if they ever feel uncomfortable about a conversation.

- **Monitor**: Monitor your child's Internet usage and monitor ALL their activity, especially when young. Check their browsing history, keep computers in a common area, and set parental controls. Just when you think they are fine, they are not.

- **Limit:** Protect access to personal information. Encourage children not to share personal information such as their full name, address, phone number, or school online—including photos with license plates, house numbers, or pictures of school uniforms. As Rania Mankarious, CEO of Houston Crimestoppers, says, "Post like a celebrity!" Celebrities know the importance of maintaining privacy for the sake of safety. Check out her book, *The Online World*, for more information.

- **Report:** If you suspect your child or another child is being groomed online, report it to the authorities. How will you know? Look for changes in behavior or new friends that have come around. Predators use peers as "bait" to get kids to let their guard down. You can contact local law enforcement or report it to the National Center for Missing and Exploited Children.

- **Advocate:** Encourage platforms to act. Advocate for social media and other online platforms to be more active in preventing grooming and other forms of online abuse. Please encourage them to implement safeguards such as age verification, content moderation, and reporting mechanisms. The same technology they use to know if someone is talking about vaccinations or Jan 6th can be used to save our children—it's a matter of will.

While I was writing this book, the US government took a stance for the safety of our children using social media. This is a good sign, and the following is the body of what was proclaimed by our government. I felt all parents and concerned family members should read this:

## Surgeon General Issues New Advisory About Effects Social Media Use Has on Youth

# Mental Health

*Surgeon General Dr. Vivek Murthy Urges Action to Ensure Social Media Environments are Healthy and Safe, as Previously-Advised National Youth Mental Health Crisis Continues.*

Today, United States Surgeon General Dr. Vivek Murthy released a new report.

While social media may offer some benefits, there are ample indicators that social media can also pose a risk of harm to the mental health and well-being of children and adolescents. Social media use by young people is nearly universal, with up to 95% of young people ages 13-17 reporting using a social media platform and more than a third saying they use social media "almost constantly."

With adolescence and childhood representing a critical stage in brain development that can make young people more vulnerable to harms from social media, the Surgeon General is issuing a call for urgent action by policymakers, technology companies, researchers, families, and young people alike to gain a better understanding of the full impact of social media use, maximize the benefits and minimize the harms of social media platforms, and create safer, healthier online environments to protect children. The Surgeon General's Advisory is a part of the Department of Health and Human Services' (HHS) ongoing efforts to support President Joe Biden's whole-of-government strategy to transform mental health care for all Americans.

"The most common question parents ask me is, 'Is social media safe for my kids?' The answer is that we don't have enough evidence to say it's safe, and in fact, there is growing evidence that social media use is associated with harm to young people's mental health," said US Surgeon General Dr. Vivek Murthy. "Children are exposed to harmful content on social media, ranging from violent and sexual content, to bullying and harassment. And for too many children, social media use is compromising their sleep and valuable in-person time with family and friends. We are in the middle of a national youth mental health crisis, and I am concerned that social media is an important driver of that crisis—one that we must urgently address."

Social media usage can become harmful depending on how much time children spend on the platforms, the content they consume or are otherwise exposed to, and how it disrupts activities essential for health, like sleep and physical activity. Importantly, different children are affected by social media in different ways, including based on cultural, historical, and socio-economic factors. Among the benefits, adolescents report that social media helps them feel more accepted (58%), like they have people who can support them through tough times (67%), like they have a place to show their creative side (71%), and more connected to what's going on in their friends' lives (80%).

However, social media use can be excessive and problematic for some children. Recent research shows that adolescents who spend more than three hours per day on social media face double the risk of experiencing poor mental health outcomes, such as symptoms of depression and anxiety; yet one 2021 survey of teenagers found that, on average, they spend 3.5 hours a day on social media. Social media may also perpetuate body dissatisfaction, disordered eating behaviors, social comparison, and low self-esteem, especially among adolescent girls. One-third or more of girls aged 11-15 say they feel "addicted" to certain social media platforms and over half of teenagers report that it would be hard to give up social media. When asked about the impact of social media on their body image, 46% of adolescents aged 13-17 said social media makes them feel worse, 40% said it makes them feel neither better nor worse, and only 14% said it makes them feel better. Additionally, 64% of adolescents are "often" or "sometimes" exposed to hate-based content through social media. Studies have also shown a relationship between social media use and poor sleep quality, reduced sleep duration, sleep difficulties, and depression among youth.

While more research is needed to determine the full impact social media use has on nearly every teenager across the country, children and adolescents don't have the luxury of waiting years until we know the full extent of social media's effects. The Surgeon General's Advisory offers recommendations stakeholders can take to help ensure children and their families have the information and tools necessary to make social media safer for children:

- **Policymakers can** take steps to strengthen safety standards and limit access in ways that make social media safer for children of all ages, better protect children's privacy, support digital and media literacy, and fund additional research.

- **Technology companies can** better and more transparently assess the impact of their products on children, share data with independent researchers to increase our collective understanding of the impacts, make design and development decisions that prioritize safety and health—including protecting children's privacy and better adhering to age minimums—and improve systems to provide effective and timely responses to complaints.

- **Parents and caregivers can** make plans in their households such as establishing tech-free zones that better foster in-person relationships, teach kids about responsible online behavior and model that behavior, and report problematic content and activity.

- **Children and adolescents can** adopt healthy practices like limiting time on platforms, blocking unwanted content, being careful about sharing personal information, and reaching out if they or a friend need help or see harassment or abuse on the platforms.

- **Researchers can** further prioritize social media and youth mental health research that can support the establishment of standards and evaluation of best practices to support children's health.

In concert with the Surgeon General's Advisory, leaders at six of the nation's medical organizations have expressed their concern about social media's effects on youth mental health:

*Social media can be a powerful tool for connection, but it can also lead to increased feelings of depression and anxiety—particularly among adolescents. Family physicians are often the first stop for parents and families concerned about the physical and emotional health of young people in their lives, and we confront the mental health crisis among youth every day. The American Academy of Family Physicians commends the Surgeon General for identifying this risk for America's youth and joins our colleagues across the health care community in equipping young people and their families with the resources necessary to live healthy, balanced lives.* —**Tochi Iroku-Malize, M.D., MPH, MBA, FAAFP, President, American Academy of Family Physicians**

*Today's children and teens do not know a world without digital technology, but the digital world wasn't built with children's healthy mental development in mind. We need an approach to help children both on and offline that meets each child where they are while also working to make the digital spaces they inhabit safer and healthier. The Surgeon General's Advisory calls for just that approach. The American Academy of Pediatrics looks forward to working with the Surgeon General and other federal leaders on Youth Mental Health and Social Media on this important work.* —**Sandy Chung, M.D., FAAP, President, American Academy of Pediatrics**

*With near universal social media use by America's young people, these apps and sites introduce profound risk and mental health harms in ways we are only now beginning to fully understand. As physicians, we see firsthand the impact of social media, particularly during adolescence—a critical period of brain development. As we grapple with the growing, but still insufficient, research and evidence in this area, we applaud the Surgeon General for issuing this important Advisory to highlight this issue and enumerate concrete steps stakeholders can take to address concerns and protect the mental health and wellbeing of children and adolescents. We continue to believe in the positive benefits of social media, but we also urge safeguards and additional study of the positive and negative biological, psychological, and social effects of social media.* —**Jack Resneck Jr., M.D., President, American Medical Association**

*The first principle of health care is to do no harm—that's the same standard we need to start holding social media platforms to. As the Surgeon General has pointed out throughout his tenure, we all have a role to play in addressing the youth mental health crisis that we now face as a nation. We have the responsibility to ensure social media keeps young people safe. And as this Surgeon General's Advisory makes clear, we as physicians and healers have a responsibility to be part of the effort to do so.* —**Saul Levin, M.D., M.P.A., CEO and Medical Director, American Psychiatric Association**

*The American Psychological Association applauds the Surgeon General's Advisory on Social Media and Youth Mental Health, affirming the use of psychological science to reach clear-eyed recommendations that will help keep our youth safe online. Psychological research shows that young people mature at different rates, with some more vulnerable than others to the content and features on many social media platforms. We support the advisory's recommendations and pledge to work with the Surgeon General's Office to help build the healthy digital envi-ronment that our kids need and deserve.* —**Arthur Evans, Jr., Ph.D., Chief Executive Officer and Executive Vice President, American Psychological Association**

*Social media use by young people is pervasive. It can help them, and all of us, live more connected lives if, and only if, the appropriate oversight, regulation and guardrails are applied. Now is the moment for policymakers, companies, and experts to come together and ensure social media is set up safety-first, to help young users grow and thrive. The Surgeon General's Advisory about the effects of social media on youth mental health issued today lays out a road map for us to do so, and it's critical that we undertake this collective effort with care and urgency to help today's youth.* —**Susan L. Polan, Ph.D., Associate Executive Director, Public Affairs and Advocacy, American Public Health Association**

The National Parent Teacher Association shared the following:

*Every parent's top priority for their child is for them to be happy, healthy, and safe. We have heard from families who say they need and want information about using social media and devices. This Advisory from the Surgeon General confirms that family engagement on this topic is vital and continues to be one of the core solutions to keeping children safe online and supporting their mental health and well-being.* —**Anna King, President of the National Parent Teacher Association.**

This is a positive step by our government, but it needs to go further. If we don't take action to protect the vulnerable, no one will. Stopping the grooming of children on the Internet is our collective responsibility. We need to work together to create a safer online environment for future generations.

# TECH ACTION

One huge step for parents: learn about your children's favorite online sites and take the time to spend time inside their world. Spend time getting to understand what your children are drawn to and only by your experiencing their favorite content can you speak to them about it. It makes sense that the smartphone generation would require parents to know the subject before just listening to them. Win by understanding what is attracting their attention and then go further and understand the addiction. Try it out! This will fundamentally change your relationship with your children and their likes.

Scan this QR code for additional information
on the topics in this chapter.

markstross.com/book-ch-6

# CHAPTER 7

# Reality vs. Virtual

*Once you have taken away reality,*
*you have taken away hope.*

Now, if the dose of reality from the last chapter seems like too much to handle, you could tune it out by escaping into a virtual world—which is precisely what Big Tech hopes you'll do. But is the exchange of reality for something virtual worth it?

If reality is based on what we see, and technology is evolving to depend more on AI, reality may not be what you see in the next five years. Imagine walking around a city or a mall; AI has created every graphic you see. If AI made it, is it real? Can you trust its intent? In the next five years, it's plausible that pictures will be created by neural networks, which is how AI works, based on observations of the world through data sources, like security cameras throughout a whole city. And yes, cameras are just about everywhere. Big Brother is for sure watching. You can visit websites today that will show you webcam footage from all over a particular city. That information can be constantly fed into computers that learn by observing the real world. Then, using a command like "look up past observations" and combining that footage with AI's version of art can bring you a very lifelike virtual world. At the root, AI is our technology learning from us to mimic what we do and, in the future, create a better version of us.

## From Kodak to Deepfakes

The idea that our technology could become more human than we are today is not something we generally want to think about. Sadly, technology can also govern itself with a better set of ethics and morality than ours, depending on who programs it. Yes, in the future, we will be observed by our creation, and it will learn from us. This is why we need to get this phase of human existence right—embedding principles, ethics, and moral rules into our AI systems to create a better world. Doing nothing should not be an option.

So, with that, let's go to the dark side of AI, starting with deepfakes, which is probably as scary as we can go. You may have heard about deepfakes, but if you haven't, a deepfake is AI technology that merges an actor's video with another person's video representation to create their likeness. So, you can have a video of someone doing just about anything and then merge it with a video of someone famous, like President Biden. The deepfake technology will put the expressions and likeness of President Biden on the video of the actor so it looks, sounds, and appears to be President Biden—but it isn't. It's a fake representation of him—a deepfake. It's so realistic that the naked eye can't tell the difference. That's because the deepfake software analyzes Biden's human organic visual and audio patterns and then takes the other person's words and actions and injects Biden's human patterns into a digital "sandwich," which makes a video that looks like President Biden speaking someone's else's lines. I think deepfake technology is scary. It can be used for very malicious purposes. But I don't want to scare you; I want to inform you. With the media hype on almost every headline, we must ask ourselves if what we see is real, especially as we move into 2030 and beyond. Can we trust the things we see online and on television? More importantly, it's disconcerting when you realize that Microsoft and huge companies like Microsoft are starting to create algorithms that will look for deepfakes in everything we do.

Now the supposition is—and I know this is a bit of a stretch—when you upload a video, the first thing that's going to happen is that it will be assumed that it is an illegal video because the first thing tech companies are going to do is analyze that video to see if it is fake. Irony?

Microsoft decided to invest in anti-deepfake technology a few years ago when it became apparent that deepfakes could disrupt their business and wreak havoc on LinkedIn and other outlets they own as they buy huge stakes in AI. This started in 2016, specifically, but believe it or not, this goes back to Kodak circa 1985. I know that may sound quite surprising, but in 1985, Kodak was playing around with digital technology, and for the first time, they went on air on ABC. They proclaimed that they could change reality by taking a digital image and superimposing a new reality on that image. From 1985 to 2023, we now have the capacity in real time to change the actual video and put new audio into a video of things that were never said. For example, I could take a video sampling of my buddy Bulldog's voice and image, download a few episodes of his morning radio show into a website that Adobe created called Voco, and in 40 minutes, Voco could start to produce whatever I type on the keyboard. I could tell virtual Bulldog anything I wanted him to say, "Mark, your *Tech Byte* show is awesome!" Within 40 minutes after sampling Bulldog's image, I now control a Bulldog facsimile. I have recreated his image. What will I do with it? I can post him telling the world to listen to my show. I can make him do anything virtually. The stakes are raised when we think of deepfakes of world leaders going out and making threatening statements or declaring war. So, we must take a moment and think about the power this deepfake technology brings to the human equation, its ability to change what people have said, and the power to change history, the future, and current events.

Since most people receive their news and information from the digital world, usually via an intelligent source using video, deepfakes can mess people up if they fall for the information conveyed through the deepfakes.

Today, more than ever, our society is experiencing "truth decay," and deepfakes aren't helping. Truth decay is the concept that a society does not trust its leaders, or the media and that politics have become highly toxic because truth, the fundamental issue, is being eroded.

The RAND Corporation, a prestigious think tank that advises the US government on policy issues, commissioned a report on truth decay, a term they coined to describe the diminishing role of facts and analysis in American public life. The report, published in 2018, explores the causes, consequences, and potential solutions for truth decay, which is characterized by four interrelated trends: increasing disagreement about facts and data, blurring of the line between opinion and reality, increasing influence of opinion and personal experience over fact, and declining trust in sources of information. The report argues that truth decay seriously threatens democracy, civil discourse, policymaking, and national security and calls for urgent action to restore the role of facts and evidence in public life.

Who knows what to believe, whether the information you receive is accurate, or if it's biased to make you genned up? Now, let's take an even closer look at this technology. Today, computers can do what is called deep learning on a subject. For example, with my hypothetical deepfake video of Bulldog, to recreate him, the computer is looking at how he paces, how he breathes, and how his intonations are. It creates a Bulldog database of what makes him different and unique. After the computer has captured all of Bulldog's nuances, they can be reproduced accurately with technology.

A deepfake can happen today with any passionate fan going to a website and sampling Bulldog's voice, then taking a whole bunch of Bulldog's photographs from all his public appearances and social channels—the more pictures, the better, including videos they found on Bulldog's friends' social feeds. Let's pretend the person, the fan, gets about 200 likenesses of Bulldog. They put all of that into a computer program with some genuine video. With all of that, they can create a virtual bulldog. To make it even better, they put sensors on themselves and make virtual Bulldog do precisely what they're doing, with Bulldog's voice, even though they speak. Depending on what they make, this could land Bulldog in jail, court, or much trouble. There's no defense for it. It's right there, in black and white. It appears that Bulldog said and did whatever was presented, and he should pay the price...except he didn't do anything.

Deepfakes did this already with Obama's image, and they got a lot of publicity for it. You can go to YouTube and look up "Obama deepfake" and see it. They also did it with George Bush, and they've done it with quite a few other celebrities. The bottom line with deepfakes is they are creating a completely alternate reality. We must treat this seriously and consider the warning. Don't assume that what you see online is gospel truth any longer. Consider your sources before you get thoroughly worked up by a video you just watched, because it may be as fake as a three-dollar bill.

I can't say it enough—it's time for people to act. It's time to go to local leaders and demand that we create new laws to handle these issues. We are not powerless; the provincial government is where we start to protect ourselves—at school board meetings, city council meetings, and going to state leaders. Asking for rules to govern how deepfakes will be handled in your community should be a priority. Think about it—what happens if a kid tries to get a teacher fired by using deepfakes? Your imagination can fill in the rest.

Sometimes, complacency is just a form of ignorance. Once people are informed, they're inspired to act. If people stay complacent during this time, we'll all wake up one day realizing we can't trust anything we see anymore.

It's funny to me that Microsoft is investing in computer AI to combat computer AI-created fakes. Now, we're going to have computers checking to see if computer-generated content is fake or not. Imagine computers policing computers. How far can you trust any of this? I want to point out here that this deepfake technology used to be the domain of scientists, the domain of a lab. Today, it's the domain of any student or kid who has time and wants to play around.

Discovering a fake video is much more complicated than just looking at it and going, "I think that's fake," especially with how good technology has become. Let's look at what happens in the music industry. People want to listen to the best version of music possible. I love hearing a master-quality cut, which is the first recorded version from the studio. The master quality is always going to be pristine. Then, if you make a copy of it, you degrade the master a little bit, and then usually, you make a distribution master from the original master. The distribution master sometimes has more cuts made from it, and usually, you're three generations away from the master for distribution in music. When it comes to digital images and videos, what you see digitally is made up of individual pixels, and the number of pixels you have determines the resolution of your photo or video. A pro can tell a lot about how an image was made by how the pixels are arranged. If I see an original deepfake video, I can probably tell that it's fake because of the telltale signs of manipulation in the way the pixels look that make up the image. But as soon as a copy is made, things start to change. If I have a master video and upload the master to YouTube or Facebook, well, those platforms have their compression algorithms, and they recompress the video. Once they've done the compression and "wizardry" on the image, it's hard to tell if it's been manipulated due to the recompression. So, for example, every time a deepfake picture is duplicated or cloned, it becomes more and more fake-looking against the master version of the same video. But the same thing happens with every real video. Every time a fake video is duplicated or uploaded somewhere, it becomes harder to determine if it's real or fake. In other words, reality diminishes every time we make a new copy of it.

Let's bring it all home. You might be thinking, *I'm not famous. I don't need to worry about this.* But people are making deepfakes of regular people all the time. What would happen to your life if a deepfake about you was released into the world? What are the protections for you as an individual? Remember, deepfakes will become more common now because it is no longer under the purview of science or academia—they are now under the purview of websites that can do it on the fly.

Google and Adobe have a huge problem, don't they? Think about it. How do you admit you're working on AI to reproduce reality and then say it won't do any harm? It's impossible to highlight the good without the harm in this situation, and I think everyone has a big question, "Well, how do you protect us against any of this?" Hopefully, as you read further, this will be explained and you feel encouraged to do your part.

## The Metaverse

We touched on this earlier, but remember, your brain doesn't know the difference between what is real and what is simulated. If you tell your brain something is real, your brain will believe you. If you experience something that triggers emotion or activates the pleasure centers of your brain (meaning you get a dopamine hit), your brain doesn't care if you can physically touch it. You don't have to. Your brain only knows that whatever is happening is triggering all the "feels!" So, here we have this glorious metaverse, where people can leave real life to enter an utterly digital fantasy world and do whatever they want. So, what could go wrong with this technology? What's wrong with a bit of escapism without a foreseeable risk? I'm so glad you asked.

Since the 1990s, computer programmers have been creating "sandbox" digital games, meaning the game consists of a whole world you interact with. It's not like a game "on rails," which takes you from one level to another level as you complete tasks. Sandbox games allow you to explore the game world without rules or parameters. These games offer adventure and strategy and have become very popular. For example, first-person shooter games are sandbox games that require players to be immersed in a new world. With virtual reality, you can immerse yourself visually inside these fantasy gaming worlds on a whole new level. In 3D, these worlds look so real that your brain thinks it is accurate. How cool is that? Well, it depends, because, from the gaming world, we move over to social worlds designed by social media companies that want you to spend your free time inside their worlds, buying and transacting with their goods and services. And this is essentially what the Metaverse is—a combination of a gaming world and a social world that is a virtual recreation of life. Sort of. Here's what happened when the first metaverse was created...

Top-notch developers spent years creating this new virtual world, and the time came to let the public in to test it in real time with real people. As soon as the virtual Metaverse became a reality, gangs immediately started forming to do mischief within the Metaverse. As quickly as this new "sterile" world was created, the stench of human foul play began to appear in tech channels. Scammers flooded into the new marketplace, not content with just making a buck in this new world here and there, but instead trying to find a way to double their wins and more by taking advantage of people. On top of that, virtual human trafficking and virtual sexual assault were rampant. So, at this point, the Metaverse had to develop rules. Security had to be created to monitor this activity. Almost instantly, when human touch entered the Metaverse, all the needs of an actual city became apparent—policing, education, a legal system, etc. It's amazing how fast humans add complexities to any system they are involved in. Even more interesting is that Facebook's Meta could not make their world appealing enough to convince the masses to jump in because they had not thought through these issues when they started. Meta eventually did address these issues, but there were consequences.

# The Great Reality Swap

Human beings are versatile and impressive because we can immediately switch from one concept to another. We can switch to a new idea even if we aren't finished with the old concept. When we switch to a new thing, we dive in headfirst and create all our own problems because we never really think things through before diving in. On the other hand, if we didn't dive in, we wouldn't have innovation. It's a double-edged sword.

Now, let's explore the idea of robots and virtual reality and what it means to exist as a human. You see, as robots take over our physical world and start doing everything we physically don't want to do (like clean our drains, do the dangerous things of war, make drone deliveries, drive our cars, clean our homes, and such), they are making it possible for humans to not have to lift our little pinkies. This is what's happening in the "real" world. At the same time, virtual worlds are being created constantly, and Google, Sony, and Apple are working on competitors to Meta. In the metaverse, they sell real estate based on the real-world model of "location, location, location," and people have bought in with real money. One tech article that got my attention read:

"Trillions of dollars are anticipated to be spent in a virtual world when we haven't even yet taken care of our own world. This is a tragic irony that we must confront and address. We cannot escape our problems by creating new ones. We cannot ignore the suffering of our fellow humans by immersing ourselves in fantasies. We cannot abandon our responsibility to the planet by colonizing a digital one. We must use the power of technology to enhance our reality, not to replace it."

I found this quote resonated with me, and I wanted to share it with you. It's from a book called *The Metaverse: How Virtual Reality Will*

*Transform Our Lives* by James Halliday. I think this is a powerful message that we should all keep in mind as we explore the possibilities of the metaverse. What do you think?

So, today, I'm asking you, which universe do you choose—the real world where you can touch things and look at a child and see their skin or go into an amusing world where it's like a 1990s video game? Virtual reality is crude, but you can go to a Justin Bieber concert and pay for tickets to watch a sub par experience because it's in a virtual world. My gosh, it's the newest thing.

Are we just diving in because it's the newest technology, or will we allow the metaverse owners to rule over our virtual real estate and create unlimited real estate to make more money? In fact, by creating virtual reality, they've now escaped all laws of the physical world, and they can create a reality that you have to pay into without any rules. I find it disturbing that now we have digital, non-tangible worlds owned by corporations, and people spend trillions of dollars on digital, non-tangible real estate. I mean, people are just buying space in servers.

As I write this book and try to educate everyone on technology and how to use it well and wisely, all the intriguing concepts I have come across have been unique. Throughout history, humanity has thrown itself "all in" to technologies without thinking it through. When America dropped two nuclear bombs, no one understood the consequences, the arms race, and everything that followed. It changed the world. They were only trying to win a war.

# Trading Real Coin for Virtual Coin

The challenges with virtual worlds extend beyond gangs and inflated virtual real estate. For example, Vortex is a virtual world that allows users to create and explore immersive environments with realistic graphics and physics. Vortex was founded by Mark Lostinberg, a visionary entrepreneur who wanted to revolutionize the online entertainment industry. Vortex has been up and running for two years, attracting millions of users worldwide. They offer a variety of features, such as social networking, gaming, education, and commerce. Vortex constantly evolves and expands, thanks to its users' and developers' feedback and contributions. It is more than just a virtual world; it is a platform for creativity and innovation.[3]

No one understood what he was doing when he created an entire cryptocurrency world. His goal was to create a virtual world around digital currency with unlimited size and assets and, therefore, an unlimited amount of financial transactions to profit from. Really, in my opinion, it is a complete grab not only to create a virtual currency but to lock people into purchasing with it. This fundamentally concerns me. When you're talking about virtual items being bought in virtual currencies, which necessitates the transfer of real wealth into virtual wealth—with no backing or guarantees—essentially, it will be a transfer of wealth from people with low incomes to the technology companies. This should concern all of us. When we look at the big picture, we have technology companies that have censored people who disagree with them and shamed people for not supporting their pet projects. In a harsh back door reality, they are trying to create the most significant transfer of wealth ever. Guess

---

3    Bing AI-generated summary of "Vortex Virtual World"

who they want to begin with? Our kids. They will get addicted to virtual worlds. It's sad that something seemingly harmless and fun can be incredibly harmful and unregulated.

I think parents especially need to start paying closer attention to this. Your kids will be drawn to this and the dopamine hit they get from viewing this stuff. Plus, it's where all their friends will want to be. We already talked about ways to help protect your kids, but one more thing you can do is go into the worlds they are drawn to and explore them yourself before giving your kids access. At least if you understand it, you can create rules around your child's usage. Parents, remember, you are in charge. Kids are ruthless negotiators with incredible fortitude and built with an ability to manipulate. Don't cave in. Work with them to create healthy boundaries they don't even know they need. You will never regret it in the long run.

## Lying Flat

I hope everyone gets this message; I am not trying to scare people, but the reality is scary all on its own. Virtual reality is becoming more mainstream, and AI has already arrived. I am not talking about something hypothetical. I'm already seeing millions of dollars in virtual sales transacted daily, from loot boxes in gaming to virtual real estate. The real world is going virtual, and its allure entices people. This progression brings us to a concept called "lying flat." When there is a lack of intellectual drive and real-world curiosity because everything is being handed out on a virtual platter, it changes people. When people don't feel valued for their opinions and contributions because they are being replaced and are no longer needed, there is a collapse of spirit. We see this in

China as the CCP forces citizens to conform to their way of thinking and being. In Western societies, we also see the suppression of opinion and censorship of thought. As I mentioned, this is truth decay and human spirit decay, too. Once you have taken away reality, you take away hope.

Lying flat is a problem in both China and the US. It's this belief people slip into that says, "Why bother? Because we have it sweet by living a minimalist life." Our young expected everything to be handed to them because my generation promised "better everything" in the future. Has anything been delivered? No, not really. Everything is messy, and technology is in the crosshairs of this mess.

Lying flat started in China as an expression to describe people giving up trying to better themselves because they understand that, under Chinese rule, their fate has been decided for them from the time they are tested as teenagers. So, the mentality is, *Why bother? We can just lay flat and enjoy the bounty of modern life's technologies and not worry about shining since it won't make a difference.* This mentality is now spreading through the US and other Western countries among youth who discovered during COVID-19 that they could stay home, work online, and then play video games without hassle. Screw the commute, work less, play more…and get away with it because there is less supervision. Today, employees in the West ask, "What can you do for me?" because pay alone does not motivate anymore; experiences and causes do. So, companies need to explain what they offer besides the core stuff. Nowadays, job hunting is like shopping for college and seeing grand institutions flaunt their climbing walls and campus entertainment centers instead of their higher goals. The employee wants their job to be as fun as college, and they are lying flat. In modern society, with instant success and failure shown on TikTok and other platforms constantly, the bar of success is being lowered, and the standard of what is considered normal is being reduced and seems to be simply going away. It's not easy for young adults to take off and become someone, but it is easy for them to be reminded that they are not someone every time they go on social media. Many folks believe, "I can't beat them, so why try," and this attitude is zapping the entrepreneurial zeal of the youth in Western countries. The tech algorithms highlight success through the number of views. People aren't unaware that there is a vast space between genius and dumbness. It's a shame that we alienate the average person by highlighting extremes. We miss the basic premise of life that there is a vast spectrum of people's abilities and a place for everyone.

Now, let's explore another dynamic. Let's say I create my virtual world, and it's my interpretation of perfect. In my virtual world, I am king. Everything I want is at my disposal. Why would I ever want to return to a dirty and brutal life? The virtual world will allow you to be popular all the time, but in the real world, that is impossible. If living in a virtual reality is more accessible, where does that leave humanity? This concept of giving people more and more dopamine-hitting experiences with no basis only makes us feel less linked to reality, and in many ways, we are losing our links to nature.

Studies have shown that since 2012, the start of smartphones, children are less likely to go on hikes in nature because they don't feel that nature is as stimulating to them as sitting inside on a device. It is shocking to think that kids will play on their devices as you drive through the Grand Canyon, and if you did not prompt them to look up, they would never look up.

Let that sink in a little.

Here's a tip: Take the devices away on nature trips and allow your children to experience something extraordinary—boredom or withdrawal from digital exposure. Guess what will eventually happen after their cries of protest rumble through their loud bodies? They will start looking around and find a connection to nature and humanity. They will discover a real sense of self that cannot be obtained via technology. They will build new neural pathways and grow their brains! Take the device away from yourself and your loved ones in nature and join the human experience!

## Fool's Meta

The biggest takeaway in all of this is to grasp that whoever owns your virtual world owns your time. And since the way you spend your time is the way you spend your life, who owns you? Maybe it's a horrible idea to have a tech giant own the universe you live in. You must be very careful before diving into an addictive pastime. Metaverses have poured billions of dollars into them to get the masses hooked.

It's a weird idea to think about a corporation being the creator, governor, and "god" of a virtual society. You buy something in your new society and then find that it has been taken away when suddenly the rules are changed by the "gods" you have surrendered to via their user agreements. You are seen as a user, a means to an end. Notice that no "human agreements" in the virtual or online world protect your human rights. They don't exist yet. I hope that after this book, things will be different. In the same way that the science fiction ideas from Star Trek have manifested, maybe human-user agreements will manifest in popular culture. In the meantime, please read and consider these meta-universes and their user agreements. Remember, you will own nothing, and the only part of the experience that will be yours is the memories created from it. Pretty trippy outcome if you ask me.

If I haven't made myself clear, please be careful about how you spend money on virtual items in a virtual world. Don't expect a monetary return. There are always people who do make significant bank on virtual deals, but it's the exception, not the rule. All pyramid schemes have a few winners and a large pool of losers. For example, buying Bitcoin has been proven to make people money one week and go broke the next, depending on the wind. There is no way to build a stable future on any fickle item. Don't just flock to a trend or jump on the bandwagon without information about the risks or how the technology is made.

Did you know, for example, that most crypto coin brands use graphic cards to make the computational hash that becomes a coin? Unless you are in the business, you probably have no idea what I'm talking about. So, let me back up. A bitcoin is created by solving a mathematical equation known as hash. As soon as the equation is solved, the equation automatically evolves and becomes more complex so it can be solved again, thus creating another bitcoin. The more complex the equation, the more physical energy is required to solve it. Solving these equations is known as mining. People are literally "mining" math equations all over the world. They solve the equation, and it evolves into something more complicated. A crypto ledger is shared worldwide that shows who is solving the current hash and what the next hash is to be solved for.

The hash is created by using electricity to power computers and graphic cards. It is not green at all. Since its value is created by solving increasingly complex mathematical algorithms, it requires more and more energy to create future coins—madness for a world demanding green energy. We are creating coins that require vast amounts of energy to produce. It's an environmental disaster mainly supported by the "green" community. They even go as far as justifying mining if the power comes from green sources like wind energy, not honestly looking at the carbon emissions generated by making them. They would not be needed without the bitcoin mining efforts.

As the movie franchise Jurassic Park so aptly conveyed, just because we can do something does not mean we should. Regarding crypto coins, I do not understand any logic to justify the massive energy requirements needed to mine them. If you're interested in cryptocurrencies, buy coins that use greener creation. More and more options are becoming available, and they will make the planet happier.

Most people don't realize that virtual currency and virtual universes are costly since they require server farms and electricity. So, anyone thinking green should be scratching their heads about this issue. Silicon Valley lectures us about our carbon footprint, but they are developing technologies that will expand theirs by thousands of times what it is today. If they put these new virtual universes online, the carbon footprint will become larger and larger as the resolution of these worlds increases because the power required will also increase. This is a huge consideration that virtually no one has openly discussed. Except now you know. Yes, I am sneakily making you aware of what you are doing so that you can do something about it! There is such hypocrisy!

To pay for these universes, tech companies will invent various ways to get you to take the plunge and give them monthly subscription revenue. There are rules around this sort of thing in the real world, but in their virtual worlds, it will be their rules and your compliance. We know that a sitting president was censored in this world, and Twitter did not censor anti-American terrorists in the 2020 elections. So, based on that information, who do you think will be censored in virtual worlds that have no constitution to protect your virtual rights, nor a government elected by you for the express purpose of protecting you?

My position is that a digital bill of rights should be created that includes the following:

1. If you choose, data should be transferable to a designated beneficiary upon your death.

2. All your data is owned by you even after death (the company does not own the data if a spouse or family member needs the data to be taken down).

3. Data that has been deleted should not be able to be brought back to be used as evidence. Tricky problem—in the past, you could throw something away, but today, when you throw something away, it can be magically brought back in an archive tool. These tools are constantly archiving everything on the web. The tricky part is these tools mean that you can never delete an embarrassing photo off the web. Ever. You really can't. Is that fair?

   There should be room to make mistakes in our lives. If everything we post is archived, we need digital forgiveness when we make a digital mistake. So, you delete an embarrassing photo, and 20 years into your future, it's brought back from an archiving service and you don't get a job because of it. Should that be legal? I propose that if you delete it, it cannot be used against you in the future. This needs to be made into law. This requires blockchain technology to be used on personal data to allow deletion to be permanent.

4. Neither government entities nor the press should be allowed to weaponize our data. Our privacy should be protected, and the

penalties for information leaks from tech companies should be severe enough to stop people from hurting individuals they disagree with. Tracking multiple mobile devices or electronic devices to create a larger net to gather information should be illegal. It's a slick way to get information no matter who opts out. For example, if someone takes a photo of you and posts it, AI can deduce it's you, and you are suddenly opted into the platform by association. That platform now has data on you whether you gave them permission or not. This should be illegal.

A family suggestion:

One quick suggestion is that a family should try to stay on the same smart platform so that they can master the security settings and make sure everyone is using the same settings. It's incredible how different it is when setting up an Android device vs. Apple. So, please keep it simple by being on the same platform, and your security should be tighter.

## Deal With Reality (Escapism)

Now, let's go to the Edge. Right now, we're on the verge of Gen Z going into virtual reality. Like Facebook changing its name to Meta and failing at Meta, we're about to go even further into a very dark place where people can go and hide and forget about all their problems and not deal with reality. Parents, please, if you love your children—and I know you do—remember they must get used to the factual. They must be able to negotiate what they're doing today in the real world and deal with disappointment, challenges, and difficulties. Gen Z is having problems with this negotiation because they no longer want to be in the real world;

they prefer to be in their addicting game or some other tech addiction because it feels better. I want parents to take a stand and say, "No, your phone will be off at night. You will go to sleep like we used to and get real sleep. You will not have your device underneath your pillow and have it on at night."

I believe these are changes every parent should make. After all the research and after all my radio shows, I believe we must help our kids. I hope this book helps many parents make some wise decisions—tough decisions to lead their family away from addiction to enjoying reality, which will prepare them for real life.

Most parents don't know that the box from your cable company or Internet provider sitting in your house is an open door to the online world, but it doesn't have to be. You can install locks on the entrance to the digital world just like you do in real life. There are also many kid-friendly software packages to protect children that can be installed on phones and computers to monitor your kid's activity. I've said it before and will say it 100 times again—never allow a child to have a phone without you monitoring them today. Think of it this way: Would you put an invitation in front of your house that says, "Anyone is welcome to come in and play with my child? The door is open; just come in!" Of course, you wouldn't. But that's what happens in the virtual world. You need locks to protect your family, just like you have a lock on your front door! Would you give your kids access to your gun rack? Of course not. Then why give your kids access to the most dangerous tools we've ever created—yes, technology can cost your child their life. Technology has already taken more teen lives than guns, anyway. Being a parent is not about friending your child, but it's about being a loving protector. That means it's okay to say "No" for their good—in fact, it's necessary.

# TECH ACTION

1. Restrict the times you and your children use smart devices.

2. Do activities that have nothing to do with technology to help clear your head and your children's heads.

3. Check the security settings of all your devices and make sure you set all your devices up with the same settings. Too often, you might have security to prevent being tracked set for three devices, but forget your personal iPad (any device) and, bam, you are sharing information with others you might not want them to have. Check all your security settings on all devices and do it at least twice a year. Settings can be changed through app updates.

Scan this QR code for additional information
on the topics in this chapter.

markstross.com/book-ch-7

# CHAPTER 8

# The Illusion of Ownership

*We own nothing, and corporations control everything.*
*They give very little back in return.*

Imagine you have worked your whole life taking photos and making memories, snapping photos of your family from birth until now, and all those photos are stored on your iPhone. It's not a far stretch. Your photos seem safe enough, don't they? By now, most of us have taken the plunge into e-books, online videos, reels, and digital music. But here's something few think about: *What happens to your digital photos, videos, e-books, and music when you pass on?* You might think you own those things—you paid for them—but you don't. I call this "the illusion of ownership." Now you see it, and now you don't! We're not the only ones asking. Recent reports indicate actor Bruce Willis was considering taking legal action against Apple to make sure he could pass on his digital music collection to his children. The story has seemingly been debunked on Twitter by Willis' girlfriend. Still, the report got consumers thinking more about what control they have over their digital purchases after they die. The answer is...none.

So, here's the scoop: When you own physical books, CDs, or DVDs, those items become part of your estate that can be left to your heirs. These tangible goods can be held, sold, or passed on. However, a similar transfer of ownership does not apply to your collection of Kindle or iTunes e-books or the many digital music albums, movies, TV shows, and videos you may have stored on your computer, mobile device, or the cloud. Many people don't realize that with most digital content, you don't own the content when you buy it. Instead, your purchase allows you to use the books or music. Most digital content cannot be passed on to heirs.

## What is a Licensing Agreement?

Because the Amazon license agreement is typical of the digital content industry, a close look at how this online retailer licenses Kindle content provides an excellent example of how the process works. Amazon states that digital content for the Kindle is "licensed, not sold, to you by the content provider." The agreement further stipulates that "you may not sell, rent, lease, distribute, broadcast, sub-license, or otherwise assign any rights to the digital content or any portion of it to any third party." In short, that means Kindle content can't be resold or left to an heir—it cannot even be given away or donated. Under the terms of the agreement, Kindle content cannot be transferred to another person in any way.

Amazon is simply one example. Similar restrictions apply when shopping at the Apple iTunes store, Barnes & Noble, and other online sellers of e-books, digital music, and movies. Each company licenses the content to the purchaser rather than selling it outright. The licensing agreements are similarly slanted toward the company's rights—giving the consumer's rights little emphasis.

None of the online stores that offer digital content have a clause regarding how to transfer media upon the owner's death. Currently, you cannot legally do so. Intellectual property laws that govern digital assets are archaic. However, most legal experts expect them to be reexamined in the future. Hopefully consumers will demand it.

For now, it's wise to make a backup copy of any e-books, videos, and digital music not protected by Digital Rights Management software that prevents copies from being made. Warning: Most e-books have built-in DRM protection, but many music files do not. To preserve your music, copy the files to a hard drive or thumb drive—any place your heirs can easily access.

After you pass away, Big Tech thinks they own your stuff, and it's hard to get them to release it to your family. Also, since the family member gave up their rights by giving "agency" privileges to the content storer, it's hard to make the case that your rights as a family member usurp the rights of the content storer. Bottom line: Better plan before a spouse's death about where you store their digital footprint.

What is a digital footprint, you ask? A digital footprint is an expression used to sum up all your digital content's exposure to the outside world. All the data stored external to yourself is your footprint; managing that data is critical. Most people do not think of themselves as chunks of data lying around in the world. Guess what—that is precisely why you must plan and think about where your data should exist and who should have the authority to access it. In most cases, you might not want anyone to access your accounts after your death. On the other hand, some services will manage your data after your death and even send out scheduled notifications and content after your death. For example, if a spouse was dying and you had a child still growing up, you could record videos released to the child at pivotal moments in that child's life. With technology, you can live beyond your death and comfort a loved one if you can afford it.

If it's important to you to pass on your digital content, ensure that your heirs know what devices and content you own and give them access to your account information, including passwords. In many cases, families already share digital devices, e-books, movies, and music using the same account. After the account holder's death, many keep using that account. Until the legal system catches up with technology and offers more precise guidelines on ownership and transference of digital content, this may be the simplest and best way to handle this situation.

## The Illusion of Deletion

Here's a fun fact—not only do you *not* own your data, but nothing can be erased once you post it online. That's right. Privacy is over. You see, it's the same thing as when you create a Facebook page and then delete it. It's never deleted because there's already been a scrubber that scrubbed that page and put it into an archive forever. Some programs monitor the Internet and make copies of all the Internet pages to archive everything that has happened in real time. The archives are intended to catch people who have changed the truth and to point out what was said or done historically.

Even on your computer, if you delete something from your hard drive, unless you write over that section immediately with new data, the data you just deleted is still available. Some programs will "undelete" that data. Programs like Eraser on the Mac will go and write over the deleted areas until there is nothing there for anyone to find. Otherwise, you are not deleting the data when you delete it on a hard drive, solid-state drive, or an old-fashioned hard drive. You're just telling the computer that an area is open. It's like a park where you planted trees; now, you will abandon it. The trees are still in the park until you plant new things.

Here's a real example of what I'm talking about: A woman named Sarah (name changed) had an old computer that she no longer used. She took it to a nearby recycling yard to dispose of it properly. However, she didn't realize that she had left her personal information, such as her social security number, bank account details, and credit card information, stored on the computer's hard drive. After Sarah left the recycling yard, an unscrupulous individual rummaging through the recycling yard found the computer and decided to take it home. Upon realizing that there was a treasure trove of personal information on the hard drive, they began to use it to steal Sarah's identity.

At first, the thief used Sarah's credit card information to make small purchases online. But as they became bolder, they began to open new credit card accounts in Sarah's name, take out loans, and even apply for a mortgage. As the months went by, Sarah started to receive notices from creditors and banks informing her of late payments, missed payments, and other financial irregularities. She was confused and distressed by what was happening but had no idea that her personal information had been compromised.

It wasn't until Sarah's credit score plummeted and she was denied a loan for a new car that she began to suspect something was wrong. She decided to check her credit report and discovered that someone had been using her personal information to open new accounts and run up massive debts. Sarah was devastated by what had happened.

She spent months trying to clear her name and repair the damage that had been done to her credit score. She also had to work with law enforcement to track down the thief, who was eventually caught and prosecuted for identity theft. In the end, Sarah learned a valuable lesson about the importance of protecting her personal information and disposing of old electronics properly. She always vowed to safeguard her digital identity and educate others about identity theft's dangers.

Unfortunately, Sarah's story is not unique. Some people in some recycling technology yards will steal your data because it's a fun pastime. It's supposed to be illegal, but it happens. I suggest to everyone if you want your data gone, the best way to get rid of data on hard drives is to do a long, low-level format before you give them away. Low-level slow formats reset drives back to empty drives, which takes a long time. Or keep the drivers as paperweights in a safe place. Think of a hard drive as this beautiful area of land. When you format a hard drive, you put down freeways and driveways so that you can find the data all over this land. That's all a format is; it puts down roads and allows you to find data in this vast area of land. That has no bearing on what's stored in that vast area of land.

If you have hard drives, take them out of laptops or computers. Either store them if you're unsure what to do or erase them. Or bash them with a hammer until you've destroyed it. They're tiny; they're not that large. Safe data is worth the hassle of cleaning up your digital life.

## Do No Evil!

I was motivated to move a lot of pixels to make the largest video boards in the world. I went after it. If you told me it was impossible, I'd say,

"No, it's possible, and I'm going to do it." We don't always see the big picture in this type of enthusiasm. Digital creators are doing so many big things in the tech world; they aren't seeing the big picture because, most of the time, they don't even know what the product will look like, let alone how it will impact society. After years of momentum, the reality is that real-world decisions have consequences stacking up for society to deal with. It's not one conspiracy but a whole bunch of enthusiasm creating new technologies. However, if our laws don't catch up soon and if we don't wise up to what we are creating with our technologies, as AI starts to focus on "human skill sets," human beings will become an endangered species. Since what we do is replaceable, ultimately, our ideas and services will not be needed to run society, moving forward. This is an ultimate shift from how civilization was built through human labor and mental effort. As we move forward in the digital age, silicon AI can better handle most of these civilization-building tasks.

I'll make it very clear: AI today is based on man-made rules, but when those rules start to be defined by a computer, then we are in a new reality. This new reality means that AI will control logistics and manufacturing and its role will be to serve man around the world. As AI proliferates around the globe, the interfacing rules will have to become global for AI to coexist, in the same way that most of the world uses the same banking system. I'm not generally into globalism, but this time I am. Everyone needs to agree to these rules, as with biological weapons and nuclear weapons; now we need to have rules for social media and how we see people's information globally.

The problem is that you have different regimes, like the CCP (Chinese Communist Party) inside China, who are interested in taking the technologies I've just described and using them for surveillance, as they already do. We would use the same technologies for entertainment. This is a challenging problem. I decided to work on this because I would like people to start to wake up to this. It's something that, unfortunately, I think we're very naive about as a country. The bottom line: When AI tries to hook up the world, which rules will be used and how do we get China, Russia, North Korea, and Iran to agree to our rules? And if everyone agrees to the rules, how will we enforce them? Will we give up more freedoms to allow peaceful coexistence with authoritarian regimes?

## Genned Up Geographically

Digital rights protect content and intellectual property. We're not ready yet with laws that can become globally enforceable. We have some rules right now, but they are in an old-fashioned format where each country does its own thing. It's as though we never thought about people traveling across the borders of different countries. We never considered people having subscription services that could let them access content anywhere. Let me explain.

Imagine that you are traveling overseas with your sweetheart, and you arrive at a destination, winding down and wanting to see your favorite show. You are subscribed to Netflix, so you get online, and everything's working for you. You start your Netflix service, but your favorite show is gone when you search. Instead, you see a whole new list of shows, none of which you were watching before or have paid to watch. Now, you can only access the list of shows that are allowed for you to watch in the location where you are. If you haven't experienced this, I'm going to guess you've never thought of it either.

What is interesting about this? This is about digital rights management, and digital rights management has regions. Those regions were set back in the days of VHS, and then culture graduated to DVDs. The entertainment industry set up these regions because they thought they would sell the rights to movies, TV shows, and music to certain regions worldwide. The European region, the North American region, and so on. By selling rights to regions, they isolated that part of the world and said, "This content can only play in that part of the world for a certain period, and when it's over, you can't see that content anymore."

Today, that is how movies and television shows are sold worldwide, still to different regions. Fast forward, and we now have a digital rights subscription for content. The unfair part is that you can only see your content where it's permissible, not wherever you are. If you live in the country where you purchased the content license, why can't your subscription go with you anywhere you go? After all, shouldn't it be based on your home residence? Shouldn't it be based on your preference if you're paying for the subscription? Some would say, "Well, just go around it with a virtual private network or VPN." Unfortunately, that is not a solution because the VPN services are now under attack by the content services, and they are playing a cat-and-mouse game. For example, someone in Europe wants to watch American shows, so they set up a VPN. However, Netflix is trying to stop that from occurring. So, the VPN works for a while, and then one night, when you want to see a show, Netflix updates its algorithm, and you can no longer use that VPN. Then, the VPN updates its algorithm to override Netflix's algorithm, and there is this constant tug of war. So that's not a solution. I don't believe it's fair for people to buy subscriptions in their country and not be able to take their digital content with them worldwide—like a little suitcase. It's time for people to be liberated to own their digital lives! That's what we all want.

So, once again, it's up to you and me to act and speak up and ask questions of our elected officials. Go to your local government leaders and say, "We need to be defended digitally!" Why are we still using systems that are 50, 60, or even 70 years old? Today, we have completely reshaped all our lives with digital subscriptions, iPhones, Android phones, and all these incredible technologies. Yet, we're still in the Stone Age with permissions, which proves my final point. We own nothing; the corporation controls everything, and they give back very little in return. How often have you gotten an email saying, "We are contacting you to protect you"? Most of the time, these companies are protecting themselves from some form of liability. For example, when Apple and Google removed Parler from their app stores, they could use hate speech to justify their actions. What is interesting is that at the same time that they used hate speech as a tool to deplatform Parler, the two giants had no issue with Twitter and all the Iran hate speech against the US being sanctioned while at the same time censoring a standing US president! It's a little unbelievable when you start to think about this issue. These companies don't use a fair system; they decide what is acceptable. They took down Parler, and the government did nothing about it. However, I am pleased that people's opinions are changing, and push back against these practices is occurring.

We need these companies to stay out of our personal space and not "parent" us with their form of "protection." It's creepy to think that Android and Apple know every app on your phone. They decide which apps go into their stores and censor what they allow according to their company bias and social stances. This means we are in a full-blown censored world governed by only two tech giants and their app stores. I don't think I want the "worldview" of these two tech giants to dictate what

I can view and use on my devices. At least for both giants, the tide is shifting, and new ways to load apps on an Apple device are coming out. Android already has different options for loading apps. But the native app stores on these devices will always be designed with the most accessible app access, and they should be open to everyone!

So, it's my life, my apps, my posts...corporations need to stay out of my life and yours too! They always say they protect us from something, but that's an illusion, too. They aren't protecting us; they are protecting themselves. I think it's time to take back the wheel and sit in the driver's seat to defend ourselves. We all deserve better.

## TECH ACTION

1. Plan to back up and safeguard your old storage drives. Either low-format them, keep them as paperweights, or smash them into oblivion.

2. Make family plans for where your pictures and media are kept and who will access your social media, backups, and other data should anything happen to you. Make sure everyone in the family is included in this plan, including your young children. Their data is just as important as yours. Everyone needs plans and to understand that protecting their data is crucial to their digital safety. Make written plans that help everyone know where to find your digital files.

3. Look at all the apps on your smart devices and inventory which ones you need on your devices. Then, when you have pruned your list of necessary apps, check all their security settings to

ensure every app is doing what you want it to do. For example, stop apps from running in the background of your smart device doing updates and tasks you are unaware of. Not many apps need access to your device all the time. I like to set up apps only to access my device when I have them open, but when I close them, they are closed to getting access to online resources.

Scan this QR code for additional information on the topics in this chapter.

markstross.com/book-ch-8

# CHAPTER 9

# The Illusion of Power— Green Power

In February 2021, a failure of people, ideas, and strategy brought Texas to its knees during a massive freeze. The issue was the power grid. They almost had to restart the whole thing. It was unprecedented. It's tough to restart an electrical grid if it ultimately loses power. Thankfully, it didn't, but it got close. The estimate was one to three weeks to get power back on the grid if it collapsed. Can you imagine the entire state of Texas without power for three weeks? I live in Texas, and it wasn't pretty for even a few days.

In Texas, green power makes up about 20% of the grid, and about 10% of what they consider "reliable energy" comes from that. Reliable energy is energy that does not go offline. So, for example, geothermal energy is regarded as green energy that is also consistent and reliable. A wind turbine is unreliable because the wind can stop blowing and the turbine will produce nothing. So, when the hard freeze hit, Texas lost about 10% of the overall energy they expected because the wind turbines froze. So now, imagine this: We put these massive wind propellers in the sky that turn so fast that if a bird or any debris hits them, they will get harmed and the propellers potentially damaged. When a propeller is damaged, it needs to be disposed of, and now we have a problem at landfills because these broken propellers are so large they can't be buried anywhere. They are made of fiberglass, which is not biodegradable. So, this "green" energy isn't so green, right off the bat. What is even more ironic with the 2021 freeze in Texas is that these massive propellers froze and got stuck, and to "unstick" them, they had to use gas-fueled helicopters to pour fossil fuels over the wind propellers! The very fuels they were trying to get away from to "go green" were required to kick-start the power grid. Without fossil fuels, we could not resume the functionality of our grid because it took fossil fuels to bring it back online. You can't make this stuff up.

Could this have been prevented? The answer is yes. This happened because the state of Texas did not consider that they would experience freezing temperatures across their whole state, so they did not buy the cold weather kits for the turbines, which would have included individual heaters. Imagine a worst-case scenario: They skipped buying the heaters for the turbines, ended up with a wholly frozen infrastructure statewide, and then had to use fossil-fueled machines to defrost and restart the grid.

This is a perfect example of why we must consider the ramifications of making everything "green" in our country. It goes back to having real experts and not special interest lobbyists advising government officials so they can access the best, most current data to make the best decisions. We must keep the brightest people in our country to help make sure these situations don't occur. We need top talent to think through these very complex problems. I'm not against using renewable energies; I'm just worried about how we implement them.

The 2021 freeze in Texas should have the rest of the country, from

California to the East Coast, strongly considering the ramifications of going green. For example, if we don't have coal and gas-burning plants available, at some point, if you have a grid failure like this, you won't have options to regain your power. If, for example, your solar panels and propellers have failed due to an overcast blizzard, what then? Let's say the storms get worse in the East; even the heaters won't stop some of the propellers from locking up. Wind propellers do not do as well in cold storms as you would think. The functionality of wind farms is usually reduced significantly. You might think the power output would go up in a storm, but it goes down. In ice storms, propeller efficiencies go down proportionately, so grids do not do well in freezing weather, and we don't have a solution for that today. In Texas, even if the propellers had heaters, they still would have lost a sizable portion of functionality as they do in the East on wind farms.

I think, with Texas, the funny part is that just a few months before this event, everyone was so pleased with how "green" Texas was, and the governor received an award for it! No one ever thought about what would happen if they had to kick-start the whole system if it failed.

Not only is wind power not always reliable, but wind farms are massive and drastically change the overall landscape. On my computer, I have a program called "Flight Simulator 2020," which trains airline pilots and shows the world photo realistically. Flying over Texas, you see windmills from the coast up to Oklahoma, and at some vantage points, you see them in every direction for miles. These turbines are changing our planet—a lot. One more thing about these turbine farms that gets under my skin: They are produced in China. What's insane about this is that if we're going to have a green industry, shouldn't we make everything ourselves in our own country? We could drive the price down further than China because we're Americans. This idea that American green power will come from China does not make America independent but more dependent on outside countries. We must do better with our green energy and ensure that green energy doesn't require old energy to get new energy functioning after a disaster.

So, the question is, what other options do we have? In my opinion, nuclear power is highly underrated. Something in the American psyche rejects atomic energy at a gut level. Many people will say they are against nuclear power because of the nuclear waste produced by force. Storing nuclear waste is challenging and a long-term storage problem due to radiation. So, you keep it permanently in the mountains or underground, and nuclear waste has a much smaller footprint than propeller waste. Have you ever considered the space needed to bury those propellers when they go out of commission? There's no recycling and no resurfacing and reusing them, either. Most people have never even thought about it. The misconceptions out there are far and wide. We must consider something like fusion or nuclear energy and overcome our stigmas. Maybe I'm partial because I once had a life-changing experience inside a nuclear silo. First, let me tell you what they are doing over in Germany. In Germany, due to the war in Ukraine, they have reversed their stance on shutting down their nuclear plants and are now going to expand their nuclear program.

Germany uses what's known as a green model for nuclear power because it is atomic power without water and the ability to melt down. It's more expensive, sure, but it's better than having humongous propeller waste that must be buried in the ground all over the country. Now, what makes Germany's nuclear model "green"? Germany has atomic plants known as dry reactions instead of wet reactions, like in the

United States. A dry reaction means you can walk right into the reactor core and not get any radiation whatsoever because the radiation is in pellets. Yes, it is possible to build safe nuclear power plants. It would be best if you made them on a small scale, not the big industrial scale that America does now. With America's current model, there can be an absolute meltdown if the system fails. Although I understand, compared to the old disasters—Chernobyl, and others—we've got to get with the program that nuclear energy can be produced safely. Germany has done it. Germany will gain independence for their grid by having more nuclear power. I think this is America's path forward before we get to epic breakthroughs like fusion reactors. In 2023, the breakthroughs in fusion are considerable, but it's still years away, and nuclear is a good, clean stopgap until we have something better.

If we follow the German model, they use ceramic pods and store them. They don't have to have the canisters, like we do today in the U.S., with rods that come out of the water. With our "wet" model, you must put the rods into a canister and then bury the canister as deep as you can with 1000 years of radiation. Germany doesn't have to do this. The pods are more expensive, and Americans dislike paying more. But I think it's beneficial to pay a little more than to deal with wind power failing on us. So, let's look at nuclear power again seriously and safely.

## My Nuclear Experience

One day, my genius mentor, Tim Jenison, invited me for a ride in his private helicopter to an unknown location near Topeka, Kansas, just for fun. At the time, I was blown away that Tim Jenison, whom I watched create a hip technology company that won Emmys for their products in television, invited *me* on his private helicopter. Because of Tim, technology was handed to me that changed the course of my life, and the story that follows may sound made up, but I assure you, it is not—you can look it up on the Internet!

To meet up with my friend and mentor for the helicopter ride that day, I had to drive towards the highest mountain peak in Topeka, Kansas, which is not very high. And up on this hill stood a considerable business building—the headquarters for NewTek. Approaching the corporate HQ building, I could see a pirate flag blowing in the breeze on the roof and a helicopter parked on its pad. NewTek was known for changing broadcast norms in those days, and their headquarters aimed to flaunt their pedigree of being "avant garde." They were evangelists, so to speak, for the future of broadcast. Tim was a visionary and saw that video technology would become more affordable and could scale to allow anyone to have the ability to create their content. At the time, Tim had a vision of the broadcast world, and it was dominated by expensive norms that consisted of millions of dollars of equipment needed to produce shows. That changed entirely because of what occurred inside the building I was approaching. Tim brought video creation costs down to the point anyone could play.

On this fine day, I meet up with Tim, and he takes me over to his helicopter and gets me buckled in. He tells me to put on my headset so we can speak over the loud noise of the engine and propeller. In a small helicopter, the cockpit is extremely noisy. After I'm settled in, he starts flipping switches, and the blades spin above my head. Soon, I feel the helicopter separating from the helicopter pad, and we are off to an unknown destination.

Things got real and fast. Tim swooped the helicopter above the trees and then dipped down into a wide irrigation ditch that went on for miles. Yes, he wanted to freak me out since I was a helicopter newbie. It scared me and thrilled me at the same time. To be flying down this ditch for a while felt like a trench in Star Wars when the TIE fighters swoop in to get the missile down the grate of the Death Star. Being in the helicopter with Tim was freaking cool, and he was trying to scare me—that was even cooler!

After a while, we were flying over flat plains and farmland. There was nothing but fields for miles in every direction. *Where are we going?* I wondered.

We landed in a field, and when Tim said, "We have arrived," I was a little confused. I obediently got out of the helicopter and started to see the tell-tale signs of where we were—a nuclear silo. A nuclear silo designed to launch a nuclear missile! At this rather unconventional destination, I had a surreal meeting with Tim.

So, I was getting a tour of this silo from its new owner. Apparently, Uncle Sam will sell anyone a nuclear silo if you agree to clean up the toxic waste within the silo. The toxic waste was from the lead batteries used to power the base if cut off from the primary electricity grid. And there could be rocket fuels and lubricants that must be cleaned up. The bottom line is that a silo is cheap, and the clean-up is a massive expense of ownership.

Tim's friend, who originally purchased the silo, had told Tim that the blast doors were stuck, and of course, to Tim, that was a challenge that had to be conquered. Tim had never fixed it nor had any dealings with blast doors and hydraulic pumps and machinery, but for Tim, this was essentially a Lego set that needed adjusting.

So, after a tour of the silo, we went to the control room, which was right out of 1963 when I was born. The switches to the calendar were literally out of the sixties. It was like being swooped back in time, such an authentic feel. Tim asked me to follow him. To get to the blast doors, we had to go into the concrete blast ramp that would direct the blast streak from a nuclear missile taking off. The stream of hot gases would be directed through the blast ramp out the blast doors. The doors were very high up the blast ramp, and the only way to get to the machinery was to climb up a large rope secured to the machinery room.

The climb up the rope was no small feat because it got steeper as you got closer to the top of the blast chamber. It was steep and high, and I was blown away to see my mentor climbing up the rope and disappearing into a small concrete room. I climbed up after him and met Tim inside the machinery room. The ceiling was so low we could not stand up; we had to crouch. Tim and I had a heartfelt discussion about the future in that room. We figured out some business issues, and ultimately, it was a life-changing event for me right then and there. It was one of the best and craziest moments of my life, and I wouldn't change a thing about it!

I love this story because you never know where you will end up; you need to think on your feet and be fully present in the moment. For me, it was in a nuclear silo. Many years later, I asked Tim about his friend with the silo. Tim said, "You are not going to believe me. The guy was arrested for making the silo a meth lab. He told us he was converting it to a residence!"

So, there you have it. There are lots of ways to minimize nuclear waste and reuse silos creatively. Of course, I'm talking about turning it into a residence, not the former. Okay, back to energy.

That story always reminds me of the effort that went into building that silo and the reverse engineering required to get it back up and functional. Everything about the building was purpose-built to do one deadly mission. Being in the machinery room for the blast doors and talking to my mentor about business issues and how to solve problems is the perfect segue into our energy issues. Nothing in life happens if you don't go out and do it. Like Tim fixing blast doors. Americans creating energy in reliable, renewable ways.

• • •

Our energy and supply chains must be designed with care. We can make tough decisions and pay a little more to make clean energy that is reliable, renewable, and abundant. Putting windmills off the coast of our country or on meadows in our heartland is not good-looking and downgrades views. We can do better and solve our energy problems if we look at all energy sources as viable, including nuclear, in my world. All sources include geothermal, hydroelectric, battery water dams, wind, solar, and experimental liquid batteries that store energy during the day and run generators at night.

The Department of Energy is working toward a 100% carbon-free power sector by 2035 to support President Biden's climate goals. One way to achieve this is to modernize the grid. That means supporting research and development efforts that can optimize power delivery and enhance resilience, implementing new interactive capabilities to allow the system to respond to change more efficiently, and unique measurements, data analytics, and models that leverage the latest scientific advancements in mathematics and computation to increase efficiency and reliability. Investment in energy infrastructure proposed in the American

Jobs Plan will create millions of jobs, advance clean energy capacity, and modernize the grid.

The one issue is that we must start agreeing on is to allow these solutions to occur, and with truth decay mixed in with old-fashioned stubbornness, you are left with America's dilemma. This country gets working when we agree; today, we don't agree on much. It will be tough to reach the government's goals. And we forget a state-wide outage or grid failure hurts us directly.

## TECH ACTION

1. Talk to your friends and politicians about responsible nuclear energy based on the German dry methods.
2. Check out the US government's stance on energy. The investments in the energy sphere make it a great sector to work in. If you have ideas on how to make renewable, clean, reliable energy, we need your help.

Scan this QR code for additional information
on the topics in this chapter.

markstross.com/book-ch-9

# CHAPTER 10

# A Special Kind of Courage

*The longing to fit in outdoes the longing for*
*protection and privacy. Let's change that.*

Growing up in England, Queen Elizabeth was my queen throughout my lifetime. She was known for her brutal frankness. She was that uncommon person who understood her place in this world and applied common sense leadership throughout her reign.

The Queen's 1957 Christmas broadcast, live from the Long Library at Sandringham, Norfolk, was a groundbreaking address. While it was the 25th anniversary of the first Christmas broadcast on the radio, it was the first ever to be televised. The broadcast was called "Happy Christmas." Here is an excerpt from her historic speech:

"My own family often gathers around to watch television, as they are at this moment. And that is how I imagine you now. I very much hope that this new medium will make my Christmas message more personal and direct. It's inevitable that I should seem a rather remote figure to many of you, a successor to the kings and queens of history, someone whose face may be familiar in newspapers and films but who never really touches your personal lives. But now, at least for a few minutes, I welcome you into a piece of my own home.

That it's possible for some of you to see me today is just another example of the speed at which things are changing all around us. Because of these changes, I'm not surprised that many people feel lost and unable to decide what to hold onto and what to discard, and how to take advantage of the new life without losing the best of the old. But it's not the new inventions which are the difficulty. The trouble is caused by unthinking people who carelessly throw away ageless ideals as if they were old and outworn machinery. They would have religion thrown aside, morality and personal and public life made meaningless, honesty counted as foolishness and self-interest set up in place of self-restraint.

At this critical moment in our history, we will certainly lose the trust and respect to the world if we just abandon those fundamental principles which guided the men and women who built the greatness of this country and Commonwealth. *Today, we need a special kind of courage, not the kind needed in battle, but a kind which makes us stand up for everything that we know is right, everything that is true and honest.* We need the kind of courage that can withstand the subtle corruption of the cynics, so that we can show the world that we are not afraid of the future. It has always been easier to hate and destroy. To build and to cherish is much more difficult."

Royal mic drop!

And this is where *you* come in. Every person on the planet has a reason to fear technology—either what it can do *to them* or what it can bring back, e.g., something they've done in the past. Suppose you think you're above the fray and potentially not vulnerable. Not to be rude, but that's entirely delusional. Aggressive hackers can target anyone unless they are off the grid and air-gapped. Hardly likely. It's not as easy as it looks in the movies. We are now in a day where AI can write code, and soon it will be competent enough to write complete programs that we will use every day—programs that will be better written than anything humans can do, and written instantly! So, we need to understand that technology can be evil or good. What is essential is that common sense guidelines and practices are used to create better outcomes since we can also be guaranteed that significant mistakes will be made along the way. Because human systems are not perfect, the irony is that AI will probably come to our rescue when we do something stupid. Or, in the classic AI answer to our human woes concerning creating an evil AI, the solution is a good AI. Nothing humans could do would react fast enough to counter an evil AI.

## A Prediction and Reality Check

How much reality do you see around you? When did you personally last do something for the good of all of humanity? And more importantly, when did you last do something good for yourself regarding setting boundaries with technology? For example, would you stop using TikTok if I present enough evidence that it is spying on you and it's evil? Would you stop if I told you your user devices and electronics are doing the same? Would you stop using your favorite social app or playing your amazing computer game? Would you stop using your favorite computer program if you knew your enemies were gathering data to use against you? Maybe a better way to put this is, *What will it take for you to amend your ways?* If you are not a user of technology, help those who are. Please share what you've learned and tell them to read this book!

Have you noticed that people self-censor their thoughts and ideas to fit in? Have you noticed doing this yourself? Do you forfeit common sense to stay part of your group? People will use TikTok because everyone else does…or at least everyone they want to think is important does. The longing to fit in must not outdo the longing for protection and privacy.

Today, private censorship comes down to when a country leans into a political position and leaves a substantial amount of its population out of the conversation by de-platforming specific thoughts, beliefs, ideas, and existence. My best example of this is the way American media handles stories about the left and right in America. The bias is clearly to the left, and anything on the right is headlined negatively the majority of the time. Anything left is slanted positively. This has been studied and confirmed.

When you look at censorship throughout history, two parties are always involved. One with extreme ideology usually tries to force its way in, and the other has the ideology they are trying to cancel. You know what is ironic? Throughout history, censorship has only shown that when the two opposing ideologies finally get scrutinized and discuss their differences openly, both sides are less extreme than the other side thinks. Both ideologies usually share over 70 percent of the same fundamentals, and there are only a few areas that neither side can agree on.

Now, let's get into some facts. Let's get into some stuff that will hopefully make you think. From 2009 to 2015, a preamble to the Twitter rules read, "Each user is responsible for their content, for his or her content, we do not actively monitor and will not censor user content, except in limited circumstances." From 2009-2015, people's content was considered their own, and Twitter stated they would not monitor people except in extreme cases. Then, in 2020, a sitting American President got censored, but Iranians, North Koreans, and enemies of America got a "pass" on their hate speech towards the US. The justification for the censorship sounded like this: "Our rules are to ensure all people can participate in the public conversation freely and safely. We look at the tweets, review them, and figure out if they violate our policy." CEO at the time, Jack Dorsey, said in response, "We do have a global leader policy. We believe it's important that people can see what these leaders are saying." Dorsey said that such tweets are eventually labeled as violating Twitter's policy but are left on the platform, and he maintained that policy is enforced consistently. Jack Dorsey felt that it was important that world leaders' political views were allowed to be seen, even if Twitter disagreed with them. Therefore, they would continue to keep those world leaders' opinions online.

So, Iran proclaiming that they want to destroy the United States or China, essentially bragging on Twitter about how they are torturing the Uyghurs, is allowed on Twitter, but not a US President. Now, on an episode of *Tech Byte*, I was having this conversation live on air with Bulldog. Right in the middle of the live radio program, the connection to my phone went dead. Yes, while I was talking about Twitter's hypocritical policies about world leaders, the censoring of the leader of the free world, and the fact that Twitter is acting more like a publisher with editorial review than a utility just passing on information in a virtual town square, my connection went down. Now, if Twitter as a publisher was operating with editorial bias, it doesn't need Federal Government protection against libel granted by Congress that protects online enterprises from being sued. The phone dying was a reminder of what occurs when someone is canceled. Your platform is dropped.

# The Government's Greatest Weakness

In 2021, The National Security Commission on Artificial Intelligence released its annual report, and it was 700 pages. Don't get too excited. I didn't read all 700 pages but skimmed through most of it. I read about 100 pages. Eric Schmidt, who co-founded Google, is the chairperson of this committee and is now devoting his time to helping America get ahead and stay ahead in technology. He is working to help us reach and keep an advantage in AI and other areas where we've fallen behind China and the rest of the world.

One of the best things that came out of this report is that it addressed an idea that I have been raising a flag about for a while—the tone-deafness of our government in the area of technology up until now, which they admitted in this report. I also would like to point out that this report would not have been made possible without John McCain, who set up the National Security Commission on Artificial Intelligence before his death.

In the report, this committee has identified one of the most significant weaknesses in our US government, which is that there are no technologists at the cabinet level of the president. No one can explain the technology to these diplomats and politicians as they negotiate new treaties and alliances significantly impacted by technology. The US Secretary of State, Blinken, made a speech recently in which he suggested that he now needs to bring technologists to his meetings so that he can keep up with state actors like China and Russia because most of the things discussed in diplomacy today are impacted by AI and technology. You can't talk about national security without talking about the Internet. You can't speak about weaponizing anything without AI as

part of the discussion. America needs a technologist representing its interests in the cabinet.

What makes me excited is that, finally, America is waking up! The American government has acknowledged that even though we are essentially two generations ahead of the rest of the world, the problem is that we're losing these two generations rapidly. We must go back to the same mentality we had in World War II and realize what is at stake if we lose the technology war to Russia and China—the two state actors developing systems to counter all our systems, from GPS to surveillance and firewalls. They are joining to create their economic zones, and we must protect ourselves. This translates into if we want an excellent economy, we must win the technology race.

We must realize that we're a small country compared to China. We have 380 million people in the US, compared to 1.2 billion people in China. That's a 4:1 ratio; frankly, when you look at their STEM students (the very smartest in science, technology, engineering, and math), they're producing 4:1 over US students.

What should America do to create a safe digital future? Well, first, America needs to collaborate. We must realize that we're not alone in facing the challenges and opportunities of digital transformation. We must work with other countries, civil society, the private sector, and UN agencies to build resilience through safe, trusted digital public infrastructure. We must invest in digital skills and literacy for our people, especially our youth, to thrive in the digital economy and society. And we must protect ourselves and our allies from cyber attacks, disinformation, and other digital harms that threaten our security and democracy.

Most people don't realize that in the US, we educate our competitors and send them back to their home countries. Why? Because of immigration policies. Yes, America allows the best talent from all over the world to come into our country to attend our fine American universities, gives them student visas, and then, after they get educated, we ask them to leave and reapply for status. We send them back to their native land. With this model, how will we compete when we have aggressive competitors who can outperform us in the number of students, population, and ultimate growth? Let the young, educated talent stay here and give us their most extraordinary years! Americans, please understand that you have decision-making power!

Ultimately, if America is going to compete in the global landscape, we must make an international effort and be headed in that direction. There are 22 countries, for example, that joined up to figure out who was responsible for the SolarWinds hack in 2020. SolarWinds is supposed to help protect businesses from hacks. I know it's ironic, but they got battered, and businesses suffered. It was discovered that the hack came from Russia. This is the most essential part of what I'm talking about. The government needs to be able to go to Russia and talk to them in technological terms and make them understand what will occur consequently. Instead of having politicians who essentially know nothing about technology talking about these issues, we need presidents and heads of state who are technocrats empowered to say, "You did this, this, and this, and if you don't shut it down right now, we're going to take our action and do XYZ." Government officials must understand technology better than just talking points! Surprisingly, Eric Schmidt was very vocal, saying, "We must have a counter, and an aggressive counter-counter to this idea that Putin would deny any involvement or say, 'We have nothing to do with it. It's not ours,' and then do nothing about it." (My paraphrase.)

At any rate, I'm grateful this report exists. I was getting a little scared because many of the principles and ideas I had been talking about for years on my radio program were being overlooked at the highest levels of government. But now they are starting to be addressed. Any American today can go and read this report for themselves. Just look up the National Security Commission on Artificial Intelligence. You can find the website very easily, and anyone can download the report. I hope it becomes clear that America must get very serious about technology.

## Action at the State and City Level

While there is much work to be done at the national level, there is also work to be done at the state and city level, and this is really where you can have a direct impact on future generations. We need more than agreements and policymaking. We need to put tools in the hands of the next generation to empower them to stay on the cutting edge of technology. One of the best ways I know to do this is for every city in America to set up what is known as a technology incubator. A technology incubator is an economic zone where a city gives tax breaks and financing to develop new ideas into businesses. In our country, we need to have a great deal of focus on digital business development. Let me give you an example. There is a student in your town who probably has an excellent idea for something; let's pretend it's a better battery. They need a place to experiment and work out their ideas. We need to find these types of people in their senior year of high school—they are the tinkerers. There's always a kid who is better than anyone else in computers! They need support and a place to create and develop their full potential. What usually happens is your town will lose that talent. They

will immediately move out to "greener pastures" because they will gravitate to the big cities and move on to where the opportunity lies. Imagine
if you could keep some of that talent in your community by creating
technology incubators. Imagine if each city made a mini business park
called a business technology incubator for your brightest young kids.
These places can be established in old warehouses, in low-rent areas,
and provide a place for students and anyone to play with technology.
That's how we will become unstoppable.

I believe strongly in incubators and was privileged to be shown the
Houston Technology Incubator at the University of Houston. It's an
essential project because those incubators support kids, and those kids
are our future. Now, China is incubating talent by forcing it. Our talent
needs to come from inspiration, which outperforms forced learning every
day of the week! With the right policy, America wins—keep the brains
and support their passions!

Another advantage is that partnerships can begin to emerge. For
example, the University of Houston can own some of the intellectual
property of all the students who utilize the incubator. Some of the
ideas might take off, and then the University of Houston can help with
patents by using their lawyers and resources, and everyone wins. I am
volunteering as an advisor right up front to America because I'd like to
do this in every town in the United States. This would help us become
unstoppable! But I still must nudge a little further.

If you are passionate about entrepreneurship and innovation, you
might be interested in setting up a business incubator in your city. A
business incubator is a workspace that offers startups and new ventures access to the necessary resources, such as office space, training,
mentorship, capital, and expertise. A business incubator can help new

businesses overcome the challenges of launching and growing their ventures and benefiting the local economy and community.

To set up a business incubator in any major city in the US, you would need to consider the following steps:

1. Identify your target market and industry. You should clearly know what kind of businesses you want to support and their specific needs and challenges. You can research your city and region's market demand, gaps, and opportunities.

2. Define your value proposition and goals. You should have a clear vision of what you want to achieve with your business incubator and how you will measure your success. You should also identify your unique selling points and competitive advantages over similar programs.

3. Find a suitable location and facility. It would be best if you looked for a space that is accessible, affordable, and adaptable to your needs. You should also consider your facility's size, layout, design, and equipment. You should ensure that your space is conducive to collaboration, learning, and innovation.

4. Secure funding and partnerships. It would be best to look for various sources of financing for your business incubator, such as grants, donations, sponsorships, fees, or equity. You should also seek partnerships with organizations providing resources, expertise, or referrals for your program, such as academic institutions, non-profit corporations, venture capital firms, or government agencies.

5. Recruit staff and mentors. You should hire qualified and experienced staff who can manage and run your business incubator.

You should also recruit mentors who can provide your participants with guidance, feedback, and connections. It would be best to look for people who share your vision and values and have relevant industry knowledge and skills.

6. Develop a curriculum and services. It would be best if you designed a comprehensive and customized curriculum that covers your participants' essential topics and skills. It would be best to offer various services supporting their development, such as networking events, workshops, pitch competitions, market research, legal advice, or access to funding.

7. Market your program and select participants. Promoting your business incubator to potential participants through various channels, such as social media, websites, newsletters, or referrals, would be best. You should also have a clear and transparent selection process that evaluates the applicants based on their fitness with your program criteria.

Incubation of talent is important because we are now competing against AI talent. Ask yourself, *Who do I want to win this arms race?* Do you want China to win the race on AI? If they dominate AI, that could change everything on a global level. The country with a natural edge in AI could field better weapons and ultimately out-think its opponents. Now, let me make this even more real for you. When I was growing up in the 1970s, it was thought that a computer chess game could not win against a human chess master. Today, in 2022, your smartphone chess game is more potent than a grandmaster. So why are we making it so hard for scientists to stay in this country, and why are we making it so hard for every major tech company, including the one I work for, to find talent?

We need to change that, and that's a significant message of this book. Let the talent in and let them stay!

I believe America will ultimately win because people still want to live in our country, and when they get educated here, they want to stay if they can get a green card. I know since I am one of those people, and it took me 15 years to immigrate to this country. Way too long.

Today, I'm working hard to give every American a road map to survive modern technology. I'm putting my heart and soul out there because we can improve in this area, and technology can serve us all better. It's no secret we create policies in America based on political motivations that depend on which party is in power. This is not the greatest way to develop long-term strategies as they can be shifted every two years. So, our digital policies need to be nonpartisan and treated as national security. Just as the government must make hard choices with how it handles technology, so do we.

## Increasing Your Understanding

- Your vote, from the village to federal elections, should now also be about the candidate's views on securing your town's digital future, which includes its Digital Bill of Rights (we will go into this later in the book) and the ethics your community will use moving into the robotic age.
- People must change the lawless and unethical cycle we find ourselves in.
- Ethics are a human constraint. Humans must be involved and participate to protect ethics.

- Cyber warfare comes into your kitchen whether you want it or not. However, with a proper defense strategy, it does not have to destroy your home.
- Every intrusion weakens countries, from meat processing plants to pipelines, and any hacked infrastructure weakens nations.

And most important: Informed people change outcomes, not technology.

# TECH ACTION

1. Make a list of the political leaders you can talk to, from local to state leaders. Go to events, speak to them about their technology positions, and educate them if needed. As stated before, get involved.
2. Get involved with programs that help create American jobs and technology. Incubators in your local city might need mentors.
3. Don't think someone else has the solution; get involved and speak the truth.

Scan this QR code for additional information
on the topics in this chapter.

markstross.com/book-ch-10

# CHAPTER 11

# Your Digital Hygiene

*Once your identity in some way
has been compromised,
you will wish you had taken the provisions.*

Just before writing this book, I participated in a self-actualization program out of Austin, Texas, called Discovery (discoveryprograms.org). In a profound exercise, they had us give medicine to critically sick people, and most of us gave all our medicine to the sick but forgot to keep any medicine for ourselves. We often make the excuse that we are too busy serving the world to serve our own lives.

So, keeping everything you just learned from the previous chapters in mind, I want you to use common sense to protect yourself from uncaring companies and governments. How? It would be best to build a "tool shed" of capabilities to care for your digital life. Here are a few tools to focus on:

## Digital Tools Everyone Should Understand:

The first tool I want to talk about is passwords. We all started using passwords around 25 years ago. At the time, I had no idea that the risk would be that hackers could hack the encryption algorithms used by those early sites. What is impressive is that any encryption from Y2K up to 2010 today can be easily hacked. Therefore, any passwords you had in those old schemes on mainstream companies like Marriott, not to pick on any company specifically, or LinkedIn, or other places, were decrypted on the dark web. You see, they didn't get the passwords when they hacked into these companies; they got the hashes.

The hashes are the pieces of code that your password is encrypted in. What the hackers do is quite impressive—they scan through all the words in the English language to go up against the hashes until they decode something that makes sense. Then, they can decrypt the hash with actual words and eventually figure out your password and everyone else's. It's incredibly wild that they can do it, but they can. The most important thing you must realize is if they defeat the hash of one password, and you're using the same passwords for multiple accounts, you will be hacked entirely, and they will get into all your accounts very quickly. My suggestion to everyone—and it's painful—make sure that every account you have has a different password. That is the ultimate security.

How many different accounts do you think you have right now with passwords? With my position, I probably have about 40 to 50 accounts. Well, if they're not easy for you to remember, and you have 40 or 50 accounts and must keep changing passwords, how do you keep up with it all?

To you, reshape the notion that 50 different passwords are difficult to remember. Once your identity has been compromised somehow, you will wish you had taken the provisions and taken the steps to keep better hygiene with your passwords. Unfortunately, yes, the best security is having different passwords at all your endpoints on the web. I think you will understand why in a minute. It's the ultimate security, and thankfully, it's getting easier. For example, many browsers will "suggest" a secure password when setting up the account. Using these passwords is a great way to protect your accounts if you have a place to store all your passwords securely. This is where it can get tricky. Like everything in life, there is always a weak link in the chain; in this case, that would be your master password list. I use a tool called e-wallet®. It stores my passwords on all devices and PCs, is encrypted, and backs itself up to the cloud. This program is my most critical security item. I protect its password and location. I don't advertise where I keep the master password, and I also have a buddy who has the password in case something happens to me. This is important. You never know what tomorrow will bring, so wherever you store and secure your password master list, ensure someone you trust also has the master password should anything happen to you.

I wouldn't be surprised if you were thinking *passwords are a pain,* and the truth is, they are. It is painful to do digital hygiene. This is a bummer topic because there's no way to do it right easily!

So first, use a program like e-wallet to lock your passwords away in a secure password-protected area. Do not leave your password in notebooks, Word documents, or files on your hard drive. Please make sure they are behind a protected area. The program that stores your passwords correctly will use encryption. That encrypted program becomes your master password that you will never give away to others except yourself, your buddy outside your house, and your spouse, but that's it. That is the guarded password.

I use LastPass for work and an e-wallet for home. LastPass is a password collection program you can subscribe to. It interjects passwords into what I need and collects and stores them. E-wallet is on my PC and iPad, secondary storage for my passwords in case the cloud is down. I have a master's and a secondary storage. You could almost say a backup to my master password list. My real point is that these programs do everything possible to keep my passwords encrypted and safe.

Using a password vault program, you commit to memorizing those passwords to the vaults. That is it. The vaults have all the other passwords. So, the amount of them is not an issue. Just make sure no one gets into your vault. So, remember one strong password and never give it out. So, how do I remember and create passwords?

I like to use a technique to assemble strong passwords as sentences I can remember. By using a sentence, I can remember the context and meaning of the whole sentence, which helps me remember it. This is important. I use, for example, things that I can remember, and then I throw in some digits and characters that make it very hard for the bad guys to decrypt. I remember a sentence or a thing, add to that some randomness, and generate a random pattern that only I remember. Now, it might seem a little complicated, but it isn't. You remember something that no one else will even understand. For example, an O becomes a zero, and E becomes a 3. These substitutions make the sentence a fortress of security, and the longer the password combined with these tricks, the better. So, for example, "I love my wife!" becomes IL0v3myWif3! That is a very tough password to break. If you know your rules of substitution, you can generate an excellent password that is easy to remember.

Okay, now let's talk about using cell phone numbers as passwords. It's convenient, but it comes with some downsides. Do you think your cell phone number is safe to hide your account information? This is

where the edge of technology recedes, the scary one. I used to give my cell phone to my bank for double authentication when they say, "Do you want to do two-step authentication?" Dual authentication means you go into your bank mobile app, sign in, and then your cell phone will chirp and send you a text. The text will say, "Oh, is that you?" It will give you a code, and you put that code back into your banking app, and off you go.

Now, let me give you a scenario that will be freaky. Imagine you've done that on enough of your accounts that a bad guy hacks your SIM card. Remember, we have the electronic SIM cards now. Meaning, I can activate my iPhone without a SIM card. I can go to a provider. This will most likely not happen in the United States as often because we have better checks and balances, but it has happened. You go into a phone store and say, "I've lost my cell phone, and I need to reactivate that number." You have enough information on you that proves your identity, and the phone store believes it's you, and then they turn on your cell phone with this number. That's how a hacker steals a number from a good guy. It's a form of identity theft, but now this is where the edge becomes apparent. Now they have your cell phone, they've got your identity, and they go to your bank app, and your bank sends a text and says, "Is that you?" Then, your bank sends out the encryption files to access your account. This crime started before the stealing of the phone.

Imagine if the hacker called you to set up this crime, which happened to a lady in the media not long ago. The hacker talks to her about the fact that she's got a problem with her banking account. She gave just a few pieces of information that you usually wouldn't think are a big deal. The lady gave the person on the phone her social security number's last four digits but no passwords. However, her hijacked virtual phone did the rest for her. All they needed was to get the bank to send them the request for a password reset that came to the hijacked phone. Are you guys following me? The cell phone became the gatekeeper to the woman's life, and they got right into her bank account, which was catastrophic.

Now, let's bring the scenario a little closer to home. Let's say your mate is mad at you and knows your cellphone password; with two-sided mobile authentication on your phone, your mate could get into your bank and other apps. The remedy for that is trying to ensure that if someone wants your cell phone, you choose an authentication app instead. Microsoft and LastPass have one, and others exist. If you've not used one yet, let me explain to you what they are. They are an application you can put on your cell phone called an authenticator, and you must use a password to get into that app. Once you're in that app, that app speaks to, let's say, the Microsoft servers, and when you get a request for two-side authentication, it authenticates you to Microsoft to identify you. Then, it sends its authentication back to the requesting app. Why is that safer? Because it's verifying through a whole chain link that it's you, and if a person has stolen your phone, they cannot get into an authentication app because it's buried behind a password or face ID. Are you following me? It's much safer. Now, Europe and other places are pushing for authentication apps to be the only way to confirm identity. I suggest, here in America, that as many people as possible stop using their cell phones as a primary form of two-sided authentication. It is dangerous. It should go to the authentication app; that's much safer.

Now, let's go into depth about storing your master password list of all your passwords. One question that has come up to me is, "Does this list not make you incredibly insecure because it's a vector point to hit and get everything?" Yes, this method does create a single point of failure, but the security is in the utility that stores your passwords; they are a vault masquerading as software. Incredibly, encryption is used to secure the software vault, and just like your cryptocurrency passwords, there are some things in life where you must have a log, like the cryptocurrency global ledgers secured through encryption. It would be best if you had a ledger of your passwords. The argument, "I'm not going to do that because I'm going to have one source of failure," does not hold up under today's reality. It's true, but only if you're silly enough to have that password in the open. That's the one password you should never write down. If you are struck down, it should be in your head and always shared with one trusted individual. If you don't have a trusted friend, hire a lawyer or professional security vault company. They do exist.

Another thing to consider, which makes having a unique password on every site easier, is every site will help you by sending you an email or some form of other reminder. If you filled in the site correctly, there should be an alternative way for them to get you on the site should you forget your password.

Ex-NSA agent and IT specialist Adam Anderson did a beautiful Ted Talk on three core negative beliefs that get people in trouble with their cybersecurity. People don't protect themselves because they think:

1. I'm not important, so cybercriminals will not be interested in me.
2. I don't have anything they want.
3. I can't stop them even if I want to. So, I'm not going to worry about it."

This idea comes out of countless mouths every year. The bottom line here is that these principal ideas are entirely false, and they're false based on the stats. I hope these stats will make you want to act.

Consider this: 48% of the workforce is in small business today. When people think they're not big enough to be the subject of cybercriminals, they are deluding themselves. Forty-nine percent of small businesses get hacked yearly, and 70% of cyber attacks focus on small businesses. Read that again. Seventy percent of cybercriminals will go after small businesses. Why? As many as 70% of employees participate in risky behaviors, allowing cybercriminals to exploit small businesses more widely than larger ones.

You might ask, *Am I protected if my small business is in the cloud only?* The answer is yes, but 60%. This is an important thing. So, 70% of cyber attacks focus on small business—and 70% of employees in small businesses don't use the best "hygiene" in their Internet and their password management. That is entirely wrong because of this next stat. This is the stat that is a real killer: 60% of businesses that get critically hacked in the small business category go out of business after six months because they've been hit by ransomware or an attack where the attacker demands money to unlock their files. Usually, small businesses must take out loans or deplete their savings, and they're driven out of business due to the financial impact of the cyber attack on the business.

So, when people say, "I'm not important," you're all important—anyone from a government job to the local donut shop. Everything we do in our society today, when we have a phone, computer, or tablet, must be protected. The costs are actual, and the staggering amount of impact that hacking has on our culture is too large. We can lessen it through our digital hygiene.

While robust passwords keep people from getting into our networks, can we keep hackers from finding us all together? Most of us use our technology from our home or office, and it can be located via a location address, called an IP address. This series of numbers is like a telephone number that your network uses to communicate with the digital world. It can be static or dynamic. Once hackers know your IP address, they know how to find you. That isn't good. One way to anonymously transact with the Internet is using a virtual private network (VPN). Many companies provide this service, too many for me to give you a recommendation; most do the job well. Check the reviews. Investing in a VPN makes your device anonymous and not bound to your home address. A VPN can make your connection come out of a public access hub in cities worldwide. That allows you to spoof or redirect your Internet source from your home URL (Internet address) to a public one in a city. VPNs use encryption to join their network and then direct your URL search request out of their publicly shared access points used by thousands of people at any given time. So, this makes it hard for the government, or anyone, to spy on you when you use a VPN. Your signal comes out of a company's Internet sprocket in a unique city, along with thousands of other user requests, and every user's originating destination is encrypted, meaning no one can read the digital signature address of where the digital file came from.

So, use a VPN if you don't want to be tracked. It's a standard course of action. It would be best to consider using VPNs when looking up information you don't want your Internet company to know about. You should be aware that every Internet company that acts like a utility must give the government access to back doors. If you need to keep information private and have concerns about people knowing what you're doing on the Internet, get a VPN subscription and use it on all your devices. VPNs can slow down your Internet a little, but it isn't noticeable for most people. The benefits far outweigh the inconveniences. Always turn off your VPN for troubleshooting issues. If you are stable before you start the VPN, you can more easily determine if it's broken or if something is wrong with your system.

In conclusion, the uncommon common sense is to use the security tools that are available to all of us. If we all use VPNs and unique passwords, we can reduce catastrophic cyber hacking. Too many folks have lost their life savings over digital hygiene issues. Don't become a victim. Know how to surf the edge without jeopardizing your life!

## TECH ACTION

1. Get yourself a VPN program and learn how to use it.
2. Set up VPN apps on all your devices.
3. If you work in a small organization, convince leadership to talk about digital hygiene to ensure you never have to pay cybercriminal extortion monies and go out of business. Any organization should talk about their hygiene.

Scan this QR code for additional information
on the topics in this chapter.

markstross.com/book-ch-11

# CHAPTER 12

# Bringing Humanity Back

*Hey you, that's right, you! It's decision time.*

Let's start with this: American technology infrastructure uses Chinese-made parts. What's the big deal about that? From our experience with China during COVID, we know that China will threaten to withhold any resource they control in order to get what they believe they deserve. With that in mind, why do cities and governments in the US still buy parts from China and countries that can withhold vital parts and take us down if they get angry with us? Does that make any sense to continually put us in such a vulnerable situation as a country? Like sending spy balloons above our heads, China does what is in its own best interest. And China knows the power they gain over us by always winning the bid with the lowest price. If America insists on purchasing our technologies based on the lowest bidder, then we surrender our autonomy to China.

We learned in 2023, there was collusion between social media giants and the government. Politics became intertwined with Big Tech in the true "American way," which is to go to the edge and then pull back when reality sets in. Mainstream media denied the possibility that the government played a role in Twitter's practices to censor information and impact an election. It took an individual, a private individual, not the government, to expose the truth about online censorship. After Elon Musk took over Twitter, we discovered that the "conspiracy theories" were indeed real.

My suggestion would be that every app our government uses or deploys by its employees should be digitally sanitized and deployed from American servers in the US. This should be without question the policy for every state and federal entity.

America knows how to self-correct. It's usually people like Elon Musk—or you and me—that step in to offer a new way of seeing and thinking that can bring new possibilities. Elon Musk taking over Twitter proved that the government got too involved in trying to create a political outcome and tried to adjust the political landscape in their favor. Some would argue they did, and others would argue they didn't. What I know is that Americans have the power to change their current reality—it just takes people who are bold enough to take action. So, are you one of those people? If not, why don't you become one? Don't just sit and complain about what's not going right. If you think, *nothing will ever change,* let me challenge you with this truth: People change things—so act! Start talking to people. Start a dialogue with your congressmen, senators, and your local town officials. Ask questions like:

- How many Chinese parts are used in your enterprise?
- Where else do you source parts? Has that country ever threatened us with resource blockades?
- Are the countries' politics in alignment with our values?
- How many parts could be problematic with cyber issues?
- Discuss why supporting Chinese logistics and infrastructure only weakens the US.

- Is the Chinese-manufactured part better than the American-made one?
- Is the quality comparable?
- How many compromises must be made to make the product work?
- Most importantly, what happens if China decides to withhold shipments or not to sell to us?
- What apps do the city workers use that are owned by hostile governments? If so, is the app essential? If not, delete the apps and replace them with home-brewed developed ones. Incubated in the good old US of A.

Now, I'm not just trying to advocate for buying American. Sometimes that's not possible. As a culture not wanting manufacturing in our back yard but wanting the benefits of it, we have de-industrialized so much that we can't produce as many parts. What I am saying is that we need to buy from morally intact countries that have ethics. That would be a good start. You see, after Clinton was president, in the late 90's, America dismantled half of its military factories that built everything from ammo to tanks. Today, with the Ukraine war, America has discovered that it is not able to sustain a long war because we cannot replenish our ammo. We have given so much ammo to Ukraine that America's stockpiles are running low on their backups, and this means that America and Europe are weakened for many years as they must replenish their war supplies used in conducting the Ukraine war. China, however, has not used up its stockpiles, and right now, it is at its most powerful point in modern history.

Because of American hubris, our country has not taken the threat of world war as seriously as it should have. Today, its shortsightedness is catching up. It's the same theme as the Internet; when will the generals and politicians deal with hard issues? If we have given up our factories to China, how do we make stuff to protect ourselves, and why do we allow China to have social media apps that spy on us through our Internet? Especially since they don't allow many of our apps on their network behind their great firewall.

The problem is that many Americans just want it easy, and they want the ability to do whatever they want; but this comes at a cost. Using TikTok has a price. China's profiling of Americans via data gathered from TikTok will have a profound effect in the future when they apply their data in their AI programs. Chinese AI can create deepfakes so real that we won't know the difference between reality and virtual fake reality. These deepfake influencers on social media will subtly change our attitudes towards communism and sway America away from freedom and liberty through social manipulation. This is not an idea I just came up with. It's the stated CCP's charter on how to establish their brand of communism throughout the world. There are already deepfake Chinese influencers online.

So, we need to ask our officials if they want a censor-free Internet? Which brings us to the next issue. Is our Internet healthy?

Most people understand very little about the Internet and the companies running it. The World Wide Web, or simply the web, is a vast network of interconnected information and services that rely on access points to function. Access points are the physical or virtual points through which users can connect to the Internet. These include devices such as computers, smartphones, and routers, as well as infrastructure such as cables, satellites, and wireless networks, both at home and in public places.

The web is a highly complex and decentralized system, which makes it both resilient and vulnerable. On the one hand, the distributed nature of the web means that it can continue to function even if some access points are compromised or go offline. This is because data and services can be rerouted through other access points, ensuring that users can still access the web even in the event of localized disruptions. However, the web is also highly dependent on access points, which means that any disruptions to these points can have significant consequences. For example, if a major undersea cable is damaged, it can cause widespread disruptions to Internet access and web services in the affected regions. Similarly, if a large-scale cyber attack targets key access points or infrastructure, it can lead to significant disruptions or even outages.

The fragility of the web when it comes to access points is further compounded by issues related to digital inequality and access. While many people around the world have access to the web through a variety of devices and access points, there are still large segments of the population that lack reliable Internet access or are unable to afford the necessary devices and services. This digital divide can exacerbate the impacts of disruptions to access points, as those who are already disadvantaged may be further cut off from critical services and information.

The vulnerability of these access points can have significant consequences for users and society as a whole. In order to ensure that the web remains a reliable and accessible resource for all, it is therefore critical to invest in measures to enhance the resilience and security of access points, as well as to address issues related to digital inequality and access.

You may be familiar with Amazon Web Services (AWS), the name for the backend servers that support Amazon. They have 1.4 million servers in all, which is more than they need to run Amazon, so they sell their surplus bandwidth as Internet services and web hosting to other businesses. Right now, on your phone, at least 2/3 of your apps have components hosted on AWS, so if AWS has a problem, a lot of businesses they support will also have a problem. And on February 28, 2017, they did. Amazon's Web Services (AWS) experienced a massive outage that affected a significant portion of the Internet, and much of the US. The outage was caused by a human error in the AWS Simple Storage Service (S3) team, which led to a widespread disruption of services. The S3 team was performing routine maintenance on the system, which included removing a small number of servers from one of the S3 subsystems. However, an AWS employee entered an incorrect command causing more servers than intended to be removed. This led to a cascading failure of multiple S3 subsystems, which in turn affected other AWS services that depend on S3, such as Elastic Compute Cloud (EC2) and Lambda. The outage affected a large number of websites and services that rely on AWS, including popular services like Netflix, Spotify, and Reddit.

The scale of the outage was so significant that it even affected Amazon's own Alexa voice assistant, which relies on AWS infrastructure. The financial impact was huge for every business relying on AWS. For example, Amazon itself reportedly lost up to $150 million during the outage, while some companies reported losses of over $1 million per hour of downtime. The incident highlighted the importance of redundancy and contingency planning for companies that rely on cloud infrastructure. Many affected companies had not implemented sufficient backup systems or alternative cloud providers, leaving them vulnerable to such outages. (Sound familiar, Google?) The AWS outage also drew attention to the increasing reliance on a small number of cloud providers to host critical infrastructure. While the benefits of cloud computing are clear, the concentration of critical infrastructure in the hands of a few providers creates a single point of failure that can have far-reaching consequences.

In response to the outage, Amazon apologized to its customers and pledged to take steps to prevent such incidents in the future. The company also introduced new measures to improve the resilience and redundancy of its systems, including better automation and more rigorous testing of changes before they are deployed.

During the last Amazon Internet outage, many people discovered that all their essential services were run on Amazon services. Access points failed badly. Many businesses found out that everything they thought was redundant in their business was still going through essentially one pipe: Amazon. There's a Microsoft pipe, the Google pipe and to some extent the Apple pipe, and after that, there are very few highways that can match those highways I just described. Ultimately, right now, there are very few choices.

While cloud computing offers many benefits, companies need to ensure that they have sufficient redundancy and backup systems in place to mitigate the risk of such incidents.

Now, when these tech giants are questioned in front of the American Congress, they usually claim that they are not that big or have that many users. They posit they could not do any harm because there is stiff competition; and they must compete globally against China and other state actors. When discussing this topic on the air, Bulldog said what we are all thinking—"I think they're huge and don't need protecting!"

Governments are slowly enacting constraints on the tech giants' overreach, but when the same government uses these services, it becomes messy. Right now, government and tech giants are really intertwined and indistinguishable from the perspective of public policy. The Twitter files "tweet" a narrative of government and social media working together to create their desired outcomes, not the democracy of ideas.

The relationship between governments and technology giants is complex and multifaceted. On the one hand, governments rely on these companies to provide essential services and infrastructure, but on the other hand, they must also regulate them to prevent overreach and pro-tect consumers. *Conflict of interest?* Indeed.

When governments are slow to act against tech giant overreaches due to their own use of these services, there are several potential solutions to be considered:

1. Increased transparency: Governments should ensure that there is transparency in their relationship with tech giants. This can involve disclosing details of contracts, services provided, and any potential conflicts of interest.

2. Create regulatory bodies: Governments can create independent regulatory bodies that are tasked with overseeing the activities of tech giants. These bodies can be staffed with experts who are knowledgeable about the technology sector and can make informed decisions about regulation.

3. Foster competition: Governments can promote competition in the tech sector by encouraging the development of alternative platforms and services. This can help to reduce the dominance of a few tech giants and provide consumers with more choices.

4. Encourage collaboration: Governments can work collaboratively with tech giants to address overreaches and develop solutions that benefit both parties. This can involve consulting with companies on policy decisions and inviting them to participate in the regulatory process.

5. Develop ethical guidelines: Governments can work with tech giants to develop ethical guidelines that promote responsible behavior and protect consumer rights. This can help to prevent overreaches and ensure that companies are held accountable for their actions.

6. Foster public debate: Governments can encourage public debate on the role of tech giants in society and the appropriate level of regulation.

Outcomes "gen up" people and when they find out that they have been lied to...well, it gets awkward. With the opening of the Twitter files, it was confirmed that the FBI was working with Twitter to censor people with the intent of influencing the election. America is a country protected by the Constitution, which protects freedom of speech. We are a constitutional republic. *If the government is censoring people, the government is dictating how we think.* If the government is dictating how people think, that is a dictatorship. We should be outraged!

If people were blocked from getting all the facts in the last election based on media and tech censorship, then did we have a fair election? Was the election biased? If activism warrants "any means possible" then American democracy is under siege from activists. Combine ruthless, reckless, and violent activism with the collusion of government, social media, and mainstream media outlets to control the narrative. This is how China and the Soviet Union both started their communistic beginnings. Both systems started with activism curtailing certain people's rights; culminating in a move from democracy to communism, resulting in people agreeing to live with a system years later that has been noted for its dehumanization effects of people. How did we get here? Are we waking up? It is mind boggling.

Eventually, people will find out the truth that they've been lured down a funnel that's been completely biased, and that their thinking has been swayed by an algorithm. When this happens, those people are going to feel a little bit betrayed and be a little bit negative about their country and all of this tech around them. Don't you think?

It's important to note that algorithms and personalized content are a ubiquitous part of our digital lives, and most people are aware that their online experiences are tailored to some degree, as outlined in the 2020 documentary *The Social Dilemma*. Additionally, it's important to recognize that algorithms and bias are not necessarily the same thing. While algorithms can be designed to prioritize certain types of content or users based on their past behavior, this doesn't necessarily mean that the content itself is biased. Bias can also arise from factors such as the sources of information being used, the language or framing of the content, or the cultural and societal values that shape our understanding of different topics.

Ultimately, it's important for individuals to be aware of the ways in which their online experiences are shaped, and to take steps to actively seek out diverse perspectives and information sources. This can help to counteract the potential for bias and ensure that people are making informed decisions based on their own thoughts and values. It is my passion to ensure that people know there is a better, safer way. We must be more conscientious about the things that we attract to ourselves, like the apps you put on your phone that create digital attraction. The apps you use attract a certain funnel of content to your phone, and that funnel of content may or may not be true and real. Most people don't want to admit that the addiction to the phone and social media is far more appealing than the desire to guard and maintain control of your thoughts.

## A Tale of Two Marks

I found a video by accident from ten years ago on *TechCrunch*. It was November 16, 2010, and there was a professor, a Ph.D. from MIT, who made a whole bunch of predictions. It did not have many views, but this video sent shivers down my spine. As I watched it, my jaw literally dropped. I went, "Wow, we should have started *Tech Byte* 10 years ago!" Now, before I tell you the story about this individual I saw on the video, I want you to imagine the following scenario:

Imagine you are in the middle of the COVID shutdown, and you lost your job. For some reading this, that won't take much imagination. But let's say you get behind on your bills, and then you can't afford your mobile services and you default on paying the bill. Months pass and you can finally pay again, but all your early-life pictures and everything on that phone is erased and gone forever because you're told that those photos on your phone were deleted "for your personal security" because after a period of your not paying your bill, the company purged the system for your protection.

No, really. This happens, and it happened to a friend of mine. No one thinks about this stuff, but it's a big deal to have all your data deleted if you can't pay for it. I understand we all need to pay for what we use, but the idea of those photographs disappearing if you don't pay means that you don't really own anything.

Now, we're going to jump forward and I'm going to tell you about this remarkable individual on the TechCrunch video. His name is Mark Davis, and he was educated at MIT. He started his life with Yahoo! then went on to Microsoft, and then he started his own consulting company. You can find Mark Davis on LinkedIn @MarkDavis. This is one guy you should check out. Mark Davis basically said on TechCrunch that if we do not create a bill of rights, ultimately, we would end up with an election where everyone would be screaming for a bill of rights, and the tech conglomerates will run an election, and no one would know the results. He predicted that in 2010, and it happened. I was blown away!

He came up with some ideas a decade ago, and it's time for them now to be brought out into the open. Mark Davis was part of the team at Microsoft that was building the Internet we use today. He saw and anticipated what the issues were going to be, but he did not have a big enough megaphone to overcome the inertia of all the excitement. Essentially, what he was pointing out is that we're starting to collect all this data on people and we're giving more power to tech companies than to governments, simply because people's lives are so embroiled inside the technology.

To bring this forward, Mark's views should have been seen by hundreds of thousands of people, if not millions. By now, we should have a digital bill of rights, and I wish we did. Personal data should not be owned by corporations, it should always be owned by the individual. Imagine if inside every picture and every piece of personal correspondence, we use the same tracking mechanism that we use for advertising, but we use it for you to be able to control your media. You put out a photograph on Facebook, LinkedIn, or Twitter, and you suddenly are very disappointed because maybe you drank too much. You took a photograph that you didn't like, and you want to take it down. Wouldn't it be cool if you owned your own photograph and you could just delete it on one device, and it is deleted on all devices? This is what it means to have digital control over your life. Well, that's what Mark suggested a decade ago, and that's what *this* Mark is suggesting today. A tale of two Marks—you should own your digital life! In addition to that, we also need personal digital ownership. This means that all data is controlled by the owner, no matter where it is on the web. The creator of the content is in control, not the big data conglomerate. For this to happen, we must get out of this digital feudalism, so to speak. If you aren't familiar with the term, feudalism is when a state, or king, or corporation owns something, such as land, and you work on their land and you manage a section of their land, but you yourself never really own anything. Well, if you think about it, that's what we've created in the digital world in unparalleled scope! We must rein this back in and give people "parcels of land" to own and work and develop, on the framework that's been developed by all these big tech giants. The tech giants will have to realize that it's "hands-off" when it comes to your personal data.

And this is why we must rethink and revamp our agreements with the tech companies. User agreements are required in the digital age, as companies seek to protect themselves legally and establish terms of service. However, there is growing concern that these agreements are often written in complex legal language and are not truly understood by users. Furthermore, the agreements are often one-sided, with companies dictating the terms of use without much input from users. To address these issues, some have proposed rethinking the concept of user agreements altogether. One suggestion is to rename them "human agreements" to emphasize that they are agreements between two parties, not just between a "user" and a company. This could encourage companies to take a more collaborative approach when drafting agreements that ensure users have a say in the terms of service.

In addition, there is the possibility of using AI to sign user agreements in the future. This could help ensure that the terms of service are truly understood by users, as AI could explain the terms in plain language and answer any questions that arise. It could also help to streamline the process of signing agreements, making it easier and more efficient for both users and companies. However, there are also potential concerns with using AI to sign user agreements. For example, there is the possibility that the AI could be biased or programmed in a way that is not transparent or fair to users. Therefore, it would be important to carefully consider the implications of using AI in this context and to ensure that appropriate safeguards are in place to protect the rights and interests of all parties involved.

With all of this in mind, I will bring us back to what I've been saying all along—we need a digital bill of rights. We need provisions to safeguard people's privacy, ensure transparency in how data is collected and used, and establish standards for the ethical use of technology.

# A Digital Bill of Rights

The idea of a digital bill of rights has been proposed by many advocates to protect individuals from the potential abuses of technology and data.

One of the key challenges in developing a digital bill of rights is balancing the need for security and protection with the need for innovation and progress. It's important to find ways to ensure that technology and data are being used for the greater good, while also allowing for the free exchange of ideas and information.

Here are some things the digital bill of rights should include:

1. The right to privacy and data protection: individuals have the right to know what data is being collected about them, how it is being used, and who has access to it.
2. The right to control one's own data: individuals have the right to control the use and sharing of their data, and to request that it be deleted or corrected.
3. The right to transparency: companies and governments must be transparent about how they collect and use data and should be required to disclose any breaches or security incidents.
4. The right to digital freedom of speech: individuals have the right to express themselves freely online without fear of censorship or retaliation.
5. The right to equal access: all individuals should have equal access to digital technology and information, regardless of their race, gender, or socio-economic status.

6. The right to digital literacy: individuals should have access to education and resources to help them understand how to protect their digital privacy and security.

Overall, a digital bill of rights could be an important step towards protecting individuals from the potential abuses of technology and data, while also promoting innovation and progress in the digital age. Which brings me to the most important concept in this book...

## The Power of Choice

Beyond our rights, we have the most important power of all—the power of choice. Protecting our humanity and treating those around us with respect and dignity starts by making a conscious decision to do so. It takes cooperation to make a community. That cooperation needs to extend to our new AI machines and robots that are slowly being deployed and replacing human jobs. Humanity and AI can cooperate with one another and create a community that helps each other. In fact, there are already many examples of such communities that exist today, such as the airline industry, Uber, and social platforms. AI can help humanity by providing us with new tools and technologies that improve our lives. For example, AI-powered healthcare systems can help doctors diagnose diseases more accurately and provide more personalized treatments to patients. AI can also help us in fields like agriculture, transportation, and energy by providing us with more efficient and sustainable solutions.

On the other hand, humans in turn can help AI in a number of ways. For example, humans can provide feedback and training data to improve the accuracy and reliability of AI systems. Humans can also ensure that AI is being used in ethical and responsible ways by setting standards and regulations. Overall, creating a community where humans and AI work together to solve problems and improve our lives is certainly possible. However, it requires us to be mindful of the potential risks and challenges that come with integrating AI into our society. We must work together to address these issues in a responsible and proactive manner.

## Can We Make a Pencil?

Now I want to introduce an amazing economist Milton Friedman, who used the making of a pencil to illustrate the benefits of free markets and individual specialization. The creation of a simple pencil, Friedman posited, involves the cooperation of countless individuals across different countries and industries, each of whom contributes their own expertise and labor to produce the final product. Consider, for example, the materials that go into making a pencil. The graphite used for the pencil's core may come from mines in Sri Lanka or Madagascar, while the wood used for the casing may come from forests in the United States or Canada. The rubber eraser may come from Indonesia or Malaysia, and the paint used to color the pencil may be a product of Germany or Japan. Each of these materials must then be processed and transformed and brought together in a very specific process. The graphite must be ground into powder, mixed with clay, and fired in a furnace to produce the pencil's core. The wood must be cut, shaped, and polished to create the casing, and the rubber must be molded and attached to the end for the eraser. Finally, the pencil must be painted, stamped with a label, and packaged for sale.

All of this requires the cooperation of countless individuals across different industries and countries, each of whom specializes in a particular aspect of the pencil's production. The miners, loggers, and farmers who provide the raw materials, the engineers and technicians who design and operate the production equipment, the factory workers who assemble and package the pencils, and the marketers and retailers who sell them—all are essential to the process. The making of a pencil beautifully illustrates the benefits of both individual specialization and the power of free markets to coordinate complex economic activities. By allowing individuals and companies to specialize in the areas where they have a comparative advantage—that is, where they can produce goods or services more efficiently than others—free markets can maximize economic efficiency and promote the welfare of society.

Moreover, the simple making of a pencil highlights the interconnectedness of the global economy and the importance of trade and cooperation among individuals and countries. As Friedman noted, the creation of a simple pencil illustrates "the miracle of the price system"—the ability of prices to coordinate the activities of millions of people across the globe in a way that maximizes economic efficiency and human welfare.

Today, this example remains relevant. However, it also highlights some of the challenges and issues that arise in the modern world. For example, the production of a pencil may also involve the exploitation of workers and the environment in certain countries, raising questions about ethical and sustainable manufacturing practices. Additionally, it highlights the potential negative impact of global supply chains on local economies and jobs, as certain industries may be outsourced to countries with lower labor costs. Moreover, it can be used to illustrate the growing importance of digital technology and data in modern economies.

Now, the way the world is moving, we must then consider how an AI robot might impact this process. To make a pencil, the AI robot would need to identify and gather the necessary raw materials mentioned: graphite, wood, rubber, and metal. It would then need to process these materials, shaping the wood into the proper form for the pencil, manufacturing the metal band that holds the eraser in place, and so on. To accomplish these tasks, the AI robot would need to be able to perform a wide range of actions, from cutting and shaping wood to molding rubber to creating a metal band using a manufacturing process like stamping or casting. It would also need to be able to coordinate its efforts with other machines or robots that might be involved in the production process. The AI robot would also need to be able to adapt to changes in the market conditions, such as fluctuations in the prices of raw materials or changes in demand for pencils. It would need to be able to adjust its production levels accordingly to ensure that it is producing the right quantity of pencils to meet the need but not exceed demand.

Overall, creating an AI robot capable of making a pencil would be a challenging and complex task, requiring advanced programming and automation capabilities. However, it is certainly possible, and there are already many examples of AI-powered manufacturing and production systems in use today. The principles of specialization, division of labor, and market coordination that are illustrated by the process of making a pencil are relevant today in every area of life. Just as the creation of a pencil involves the cooperation of countless individuals and industries, the production of digital technology and data involves the collaboration of numerous companies, developers, and innovators around the world.

So, the ultimate question is, *can we make a pencil today?* Of course, I'm no longer talking about a pencil you hold in your hand. I'm talking about using the metaphor of making a pencil utilizing the principles of specialization and market cooperation necessary to accomplish a goal. Society needs to create agreements and processes globally and with AI, designed in a way that is sustainable, equitable, and responsible, and that takes into account the unique challenges and opportunities of our modern economy.

As we look toward the future of AI and technology, it's clear that there are many challenges and risks that must be addressed if we hope to avoid a dystopian future. From concerns about job displacement and income inequality to fears about the potential misuse of AI and the erosion of privacy and civil liberties, there are many potential pitfalls on the road ahead.

However, despite these challenges, I firmly believe that AI and technology have the potential to be a force for good in our world. With the right policies, regulations, and ethical frameworks in place, we can harness the power of these technologies to solve some of our most pressing global problems, from climate change to global health to social inequality.

The key to unlocking this potential lies in making tough decisions today, before it's too late. We must recognize that AI and technology are neither inherently good nor evil, but rather they are tools that can be used for either purpose. We must be thoughtful and deliberate in our approach to regulating these technologies, striking a careful balance between promoting innovation and protecting the public good. This will require collaboration and cooperation across governments, industries, and academic institutions, as we work together to develop standards that ensure the responsible development and use of AI and technology. It will also require a commitment to lifelong learning and education, as we equip ourselves with the skill and knowledge necessary to thrive in a rapidly changing world.

In the end, the choice is ours. We can choose to let fear and uncertainty guide our actions, or we can choose to embrace the potential of AI and technology and work together to build a brighter, more equitable future for all. Technology doesn't have to kill the human spirit. The road ahead may be long and difficult, but with determination, vision, and a commitment to responsible innovation, I believe we can turn the threats of AI and technology into a utopia of opportunity and progress, and we can save our humanity in the process—and create some "killer" tech!

## TECH ACTION: CONNECT WITH ME

 @mstross

 linkedin.com/in/markstross

 @strossmark

 @markstross

www.markstross.com

Scan this QR code for additional information
on the topics in this chapter.

markstross.com/book-ch-12

# The Making of Killer Tech: Mark's Back Story

It was the summer of 1980, and I was driving down Benedict Canyon Drive in Beverly Hills in my Subaru XT Turbo with hydraulics. I had just graduated high school, but by then, I had experienced more than most, for which I'm grateful. I mentioned in the introduction that my mother was a well-known British actress and my father a film producer. I lived in a neighborhood with Hollywood legends. As I drove toward Sunset Boulevard, to my left was the iconic Beverly Hills Hotel, a place where glamorous and spectacular events were ordinary. I proceeded down Sunset through the hills, looking at all the enormous homes flanking the renowned route to Hollywood. Moments later, my journey transitioned from the posh, serene hills to the wildly bright and interactive Sunset Strip, filled with billboards and home to Tower Records and famous clubs like the Roxy and Whiskey A Go-Go. I had been entertained by bands at those clubs and enjoyed Tower Records, always dreaming of what I would one day do to get on one of those cool billboards. At the time, I didn't realize I would become the sign maker of the future, the digital "puppet master" for experiential realities in the most iconic places in the world. I always saw myself following in my father's footsteps as someone who gets his vision made, so I started at the prestigious and innovative theater program at Pepperdine University.

• • •

"I would like to direct a one-act play," I said, thrusting a script into the hands of Dr. Henderson, dean of theater. I was a freshman, having just graduated from high school. Dr Henderson looked at me and smiled, "Do you understand that seniors wait years to be able to do this?" I wasn't about to take no for an answer. I said, "I have the notes ready, and it is just as valid to give a freshman with talent and ability a shot at the big stage as any other classman." Dr. Henderson wasn't accustomed to getting push back from students. I got his attention because he granted me my wish. I was directing a Ray Bradbury one-act play called *To the Chicago Abyss* at the Smother Theater at Pepperdine. (Thank you, Dr. Henderson; you got the show started.) Now, it just so happened that Ray Bradbury himself was scheduled to give a lecture that fall at Pepperdine, which was four months before the opening of my one-act play. I was excited to hear him in person, and on the evening of his lecture, I arrived more than an hour early to the lecture hall, where a huge line was already forming. I was at the front of the line as the doors opened and took a seat right in front, hoping to get some time with Mr. Bradbury after the lecture. At the end of his presentation, as expected, he invited students to come to the front and speak personally with him. When it was my turn, I introduced myself to him and said, "Mr. Bradbury, I am going to direct *To the Chicago Abyss*."

"Oh really, why did you choose that one?" He looked intrigued.

"Well, it's a play we did in high school, and I fell in love with its sim-plicity. It has a huge message about not taking life for granted."

"Mark, that is one of my favorite one-act plays to direct." He paused for a moment. "Would you like my original notes from when I directed it?"

I was stunned. "Yes! Please!"

"Here is my number. Please call me after the Thanksgiving break is over." He discreetly gave me his number, and I walked away on cloud nine. I couldn't wait to call him after the holiday. I want to give a shout-out to Joe Culp, who directed and acted in this one-act during high school. I loved the play and his performance. I thought it would be a great play for my debut as a director.

Over the Thanksgiving holiday, my parents took me to a resort in Palm Springs. I was hanging out with my dad in the jacuzzi at our bun-galow when my dad looked up at a man walking by and said, "Oh hi, Ray!" To my surprise, it was none other than Ray Bradbury! He joined us for conversation, and I ended up to spending the holiday break with him, and we talked about his play. He came to my opening at Pepperdine's Smother Theatre, which gave me points with Dr. Henderson. I was sure my career in the entertainment industry was well underway, or was it?

After that experience, Pepperdine didn't inspire me the way I hoped. I transferred to the Art Center College of Design in Pasadena, CA where I knew I would be with the world's best artists and industrial designers. This was where the up-and-coming talent was, and I wanted to be personally challenged.

Art Center was a completely different animal than Pepperdine. I could describe it with one word: "intense." At Art Center, I produced a student film called *Rat Race* about a man who was a reformed molester.

He did prison time for his crime, and the moral question of the film was, *Should he be allowed to have a relationship?* I was being graded on the production, and my instructor had different ideas from me about the way I should edit my film. There was a scene where the main character had to attend some counseling sessions, and I chose an unconventional camera angle for these scenes. My professor didn't like what I chose, and we got into a dispute over it. The conflict escalated to the president's office, and he asked why I was so resistant to cutting the film as my professor advised. I held my ground, stating that editing the film my way was my "artistic view of the world," and was as valid as any other perspective. He finally acquiesced, and I ended up with my version of the film. Again, pushing back paid off.

Art Center would often have critics from the *Hollywood Reporter* come and critique the student's films, and as it turned out, they were coming to see mine. Knowing this made me both excited and nervous, as anyone would be. The morning after my premiere, I ran out to the student bulletin board where the student film reviews were posted. It was good, wow! The reporter highlighted the extremely sterile feeling created by the camera shots I chose for the counseling scenes—my precise goal and intention! The exact scenes that caused conflict in school also gave me critical success through a professional assessment. That "moment" taught me to always hold the line for what I believe in.

While at Art Center in the mid-80s, I also took classes in 3D animation using state-of-the-art tools. Back then, there were no keyboards or mice. It's hard to imagine a world with computers and no keyboards or mice but it's true. To create a computer program, code had to be handwritten on paper and then manually transferred to punch cards, using a typewriter-like device to punch the holes in the cards. The cards were then inserted into a card reader connected to the computer and scanned to see which input needed to be selected on the computer. It was quite an ordeal and extremely costly. My classmates and I wrote a 30-second, three-dimensional pixel plotting code on punch cards. Then Caltech donated $10 million of computer time to render the animation on its Cray, the supercomputer. Yes, it cost 10 million dollars to produce 30 seconds of digital animation. This was back when they were still hand-drawing animation, and for good reason. We produced extremely crude animation, and we were so proud of it that we submitted for the prestigious Clio Award in advertising. It was the first 3D animation ever submitted, and we won!

Today, the same process that took millions of dollars to create in the 80s could be performed on any smartphone in a matter of minutes. Incredible!

After Art Center, I went into television production, still following in my father's footsteps. I decided the least expensive way to get started in the industry was to do the production work myself. Yep. My father's footsteps. Digital animation was an up-and-coming technology and the topic of conversation at an exclusive tech conference I attended. The conference was held in a hotel convention center built in the 60s with low ceilings, too many columns to see booths well, and terrible air conditioning. But I assure you, no one in the room cared about any of it that day. They were too focused on the tech and the tech titans assembled. This convention brought together in one room all the great minds of technology, such as Bill Gates, Steve Jobs, and a team from NewTek. This was also the start of the personal computer age, and these "titans of technology" were all accessible to normal folks like me at that convention. So, yes, we all talked and shared. We were all considered peers in technology at the time. Of course, trying to talk to Steve Jobs at a convention 20 years later would be next to impossible!

At the convention, all the latest and greatest technologies were demonstrated. One product caught my eye, the Digiview by NewTek, a color scanner operated by using a black and white camera and a color wheel. This setup could show a possible 4096 colors—completely unheard of in the industry back then. Most computers at that time had 8 or 16 colors. Over 4000 was amazing and beyond comprehension. One caveat is that it only worked specifically with a Commodore Amiga computer. After watching the Digiview demo, I purchased one for myself along with an Amiga computer. I immediately went home and started "playing" with computer graphics. I was eager to enter the industry and make a name for myself.

Because I had purchased a Digiview, I received the manufacturer's magazines and product brochures in the mail where I read about another new product that could turn a computer into a video mixer. This was called the Video Toaster. The Video Toaster is what's known as a video switcher today. In the days before television switchers, if you wanted more than one camera angle in a video, you would set up two cameras or more and record the whole scene on every camera on film. Then, in post-production, you would edit, cut, and splice all the content to create the finished product. With a switcher, you can switch between cameras in real time while recording, saving a ton of editing time and cost. This allows you to do live shows with multiple camera angles in real-time.

The Video Toaster went from 4096 colors to broadcast quality using up to 16 million colors, and that meant the cost to produce broadcast-quality content had just come down from millions to about $3,000 in the 1990's. So, I invested in the Video Toaster, and, all in, my start-up equipment set me back about $20,000. This was the birth of my first company, Toaster Marmalade. *I mean, if you are going to have "toast" where I come from, you must have marmalade, of course!* That one purchase started my professional career, and what I did with the Video Toaster is a book in itself.

Toaster Marmalade's first paying jobs were graphics for commercials and business projects. My business partner, Jason Norris, was an Amiga technician whom I met at the Creative Amiga Computer Store in LA. The Amiga computer was the first consumer computer to do broadcast-quality work, and Jason supported my computer needs then. We started with an eclectic list of clientele ranging from the National Tattoo Convention to the American Bible Society and everything you can imagine. Years later, we didn't have the credentials that a "normal" video company would have, but we had grit and determination, which opened doors for us.

In the early 90s, we had the opportunity to work on an HBO movie called *Afterburn*. The film was about the F16 wire chafing problem in the late 80s. If you aren't familiar with that problem, just know it was not good. Numerous military planes had wiring problems, which caused them to literally drop out of the sky. It was a controversial subject matter for the Air Force, so they refused to help the production company. They didn't want to shed light on this problem, let alone make an entire movie about it. They refused to provide any military aircraft for making the film, which normally they would for movies that put them in a good light. The production company had to build mock-up planes and cockpits to create the film, but they couldn't fly them. This is where we got involved in the story. With our Video Toaster, Jason, the engineering guru, created a digital heads-up display for the F16 plane so that with a green screen and our technology, it looked like the pilot was flying. We had so much fun with this project. After finishing this job, we couldn't imagine what could top this experience. Until one day, the phone rang.

"Hello, I'm Mickey Mouse! Is this Mark Stross?"

It was Imagineering, the company behind the Walt Disney film genius! It would be anyone's dream in the industry to get this call. They wanted to meet with us right away, and we gladly obliged. The crew at Disney was told by someone who worked with us that our company "thought differently" about constructing 3D elements. Disney was in a bind with one of their rides at Epcot Center in Orlando and needed an out-of-the-box solution. Disney had an impossible two-week deadline for a vital scene they were working on for an attraction called "Ride the AT&T Network." The attraction had a motion simulator pod you could get into and "ride" a fiber-animated optical cable from one point to another. The scene had failed with traditional 3D animation, and Disney realized they would not meet their opening deadline without a fast solution. With the Video Toaster, we were up to the challenge! We skipped Christmas to do the impossible. We got the job done and then enjoyed the New Year with a huge check from Disney—our largest payday ever. The ride was at Epcot until around the mid-2010s. This whole experience taught us that sacrifice is a big part of creation and when opportunity calls, answer the phone.

Once we completed this project, things got interesting. We were then offered other projects with Disney. Up to this point, all video boards were amber in color. We developed the first content server for running true color LED video boards in the United States for Disney's Wide World of Sports Arena. We also ran it from an automated touchscreen version of the Video Toaster in real time. Then we added the first sensors that could change the brightness of the video board on the fly. None of this had ever been done before. We renamed our company to Playable Television.

Jason, who deserves a whole chapter describing his exploits, came up with the back-end technologies to make the Video Toaster a touch-screen product. This set us light years ahead of our competition and opened many doors for us. In addition, he created a product called "Blackie" that allowed us to control the brightness and color that a video board would put out depending on outdoor conditions. This set us up to be leaders in the LED industry.

From there, we took our product Blackie and our knowledge of animation around the world and had interesting meetings with people so highly influential and ultra-wealthy that no one had heard of them. It was surreal, and we were making our mark on the industry.

Another lesson I learned from my parents is always to create relationships with everyone on your life's journey because you never know who will open the next door for you. This might sound like a cliché, but it's true. Networking has been the framework for how I've succeeded in many cases. Some inventors think their ideas will bring people together, but the truth is people bring ideas together. People make projects happen, not ideas.

During this period, I went to as many conventions as I could, which is where I struck up a relationship with the team at Panasonic. They were the ones who helped me get my first major gig, making television shows when I was just out of college.

Seemingly out of the blue, one day, we got a call inviting us to meet with the team at Universal Studios. This is where I met Todd Stevens, the line producer of *Major Dad*, a TV sitcom popular in the 90s. Todd later produced a hit TV show you may be familiar with—*Friends*. Todd and I were discussing ways we could work together, and during this meeting, he asked me to meet the execs of MTM, an arm of Universal TV, so we could pitch our capabilities. In that meeting, the executive producer brought Todd and me into a side conference room and grilled us on how she thought we could never produce a television show for a fraction of the cost we had claimed. She said, "All television shows use $1 million in post equipment." We told her we could do it for $30,000 an episode. She insisted there was no way we could do special effects at the level required for that kind of money.

We left the meeting, and I immediately got to work to prove her wrong. I produced a 5-minute video personally for her using all the special effect elements in the programming that she claimed I couldn't do. It was a scrolling thank-you letter, complete with a delivery of digital roses.

She received my video about 5 hours later; she called Todd two weeks later and said, "Your stunt paid off. You got the series." Todd replied, "I don't know what I did, but I'll take it!" He called me, and we both laughed and celebrated our victory.

Our first low-budget television show was called, *Mysteries from Beyond the Other Dominion*; it was produced literally out of a converted garage. For under $30,000, we produced shows at a level that formerly required million-dollar post-production facilities. The Video Toaster made it possible, and it worked just as well for a lot less!

Around the mid-1990s, I got a call from the Cleveland Playhouse Square and was asked to fly out to Cleveland to talk to them about using our Blackie product and maybe our control system. We got the gig and ended up controlling and color-adjusting the video boards in Cleveland Square in real time. What an amazing moment for Jason and me to stand in the middle of Playhouse Square in downtown Cleveland, looking at our video boards lit up with our technology running them. We didn't understand it then, but that moment marked the beginning of a new direction for our future that brought us to where we are today.

## Stumbling Onto the Red Carpet

After we got the video boards running in Cleveland Playhouse Square, we returned to Los Angeles. In LA, I was contacted by the folks over at Planet Hollywood; this started another adventure in the technology business.

Planet Hollywood contracted us to create listening stations for each table at their restaurants, harking back to the days of the jukebox at the coffee shops in the 50s. Planet Hollywood wanted to add music videos to the mix. With our technology, you could select your music video at the table, and it would play on television screens all around the restaurant. We created listening and video streaming stations well before the iPod or "streaming" ever was a thing! This was in the early 2000s, and technology barely had the power to move video, so our video listening stations had to be custom-made with the most powerful graphics, but in a small package. We created something like a large iPod (for real), which was put into the restaurants in Orlando, Florida, and Times Square in New York.

Robert Earl, the CEO, was our contact, and he invited us to his restaurant to present our prototype to him and his investors. Usually, this would not be a problem; however, when working with Planet Hollywood, nothing ever went as planned. We had scheduled the meeting for midweek, but when the time came, we were still waiting for a few parts we had ordered, so we had to reschedule. Robert knew we would receive our final parts late afternoon that Friday and suggested we assemble the prototype on Saturday and meet with him later that day. He was extremely eager to get this project working and seen by his investors. We didn't usually work on Saturdays, but given the size of the opportunity, we agreed.

The warehouse where we were building our prototype did not have air conditioning during weekends. It was a hot Orlando summer day—generally hotter than a summer day just about anywhere else. When we assembled the parts, Jason and I were smelly and drenched in sweat. We needed to shower badly, but we didn't have time. Robert asked us to immediately bring the new units to a "small gathering" at the Orlando Planet Hollywood, which is a massive building in the shape of an actual planet.

Jason and I arrived at the parking lot of Planet Hollywood, and surprisingly, it was filled with cars. We thought, *this wasn't a small gathering!* We walked up to the building and were greeted with spotlights, a giant red carpet, the press, and paparazzi! Here we were, two dirty, smelly engineers with our prototype listening station in hand. We were ushered down the red carpet in front of the press, cameras flashing and searchlights beaming, with our machine, which had wires and electrical hookups dangling from it. I have no idea what the press thought we had in our hands, let alone what they thought of these two long-haired, sweaty nerds in grungy t-shirts walking the red carpet. We must have been a strange site to behold!

We arrived at the end of the carpet, and staff walked us to meet with Robert in a private room with a bar and barstools. We saw the back of a glamorous blonde sitting on a stool waiting for us—the investor. When she turned around, it was none other than Britney Spears. Robert was hosting a birthday party for Britney's manager, and he wanted her to see her music via our listening station and video boards. We nervously set up our prototype, and miraculously, everything went off without a hitch. The prototype was approved and deployed in Orlando and Times Square. When they launched in New York, we got to walk the red carpet again for the opening party, but this time, we were showered and dressed appropriately. That event drew all the celebrities of the day—Sylvester Stallone, Britney Spears, Bruce Willis, and many other household names. Jason and I got to party with them!

Robert was impressed by our work and allowed us more latitude in our creativity, and we came up with a project we called Planet TV. This was the local station played inside Planet Hollywood and was conceived and produced by my team. Everything was going great...until Robert went bankrupt. That hurt us financially because we invested so much in the TV show that we thought would bring us a great return. However, if our business with Planet Hollywood had not ended, there would have been no ANC and all the subsequent video board adventures we would have. We learned a lot while working with Planet Hollywood, and I am so grateful for that experience and learning from the failure of the experience!

There was always something brewing in the early years. I was also becoming more fluent with computers and their design. With our digital tools established and proven, a friend suggested that we hook up our computers to LED video boards to see how it would impact the quality of the video boards. They suspected we could make the video boards look much better than they ever had before. But how would we do that? Who would allow us to experiment with our equipment on million-dollar

LED video boards? Well, we got creative.

One night, like a bunch of frat boys on a pledge, we sneaked into the San Francisco Convention Center after hours. Well, not completely without authorization. Remember that bit about networking I mentioned earlier? Some folks watching our success thought we had a pretty good "mouse trap" and wanted to give us a chance with their LED boards. So, the head salesperson for SACO LED video screens let us into the San Francisco Convention Center without SACO corporate knowledge. What happened that night changed my life and the industry forever. My server made the video board look amazing and added dissolving video content that blew everyone away—especially the SACO executive who arrived the next morning to do the trade show and found their board displaying footage with dissolves better than anything he had ever seen before!

I had one prototype, and suddenly people wanted more. I was beginning to grasp just how revolutionary our technology was. The content servers were in demand after this first test, and I will always be grateful for the people who took the risk to sneak me into the trade show that night.

Let me encourage you to take a chance on a new idea. It requires risk to move forward. People who embrace risk and change move the needle every time!

Playable Television was about making content easy to use on touch-screen servers. It was a concept ahead of its time because of the prior expense. Yes, before streaming and smartphones, we played with real-time video that could dissolve into new video content immediately, without delay in broadcast quality. My technology, created by three people and some off-the-shelf equipment, was already faster than the interactive broadcast devices from the tech giant, the Philips Corporation—a $20 billion company with 80,000 employees throughout 100 countries.

Every technology we built was always faster than our competitors. The competition, like Philips CDTV, had a 20-second delay on average when transitioning from one interactive scene to another. We could transition in real time, so the difference was significant. In the 1990s, this was a big deal. This technology was the core idea behind my work with ANC, ushering in a new wave of visual technologies for professional sports stadiums around the country and then expanding to larger places throughout America.

Over my career, one thing has been a constant: when you create, sometimes you ruffle feathers and even straight-up make people angry. If you have ever been in a room where everyone else does not get it, then you know a little about what it is like to be me. You don't cave when people don't understand you; you refine your ideas and continue to make progress, bringing everyone with you along for the ride.

## The Ultimate Test

When we created our prototype of Playable Television, we were excited about its potential. After we debuted it at The National Association of Broadcasters (NAB), we were invited to set up a demo in the Philips booth at Infocom, another multimedia convention, around 1993.

Philips was demonstrating the first interactive music video, which happened to be "No World Order" by Todd Rundgren, a popular rock-and-roll artist. You could manipulate the video with touchscreen buttons, which was quite innovative then. You could completely arrange the whole album to your liking with dozens of customization points to change the content.

We set up demo stations for our technology in the Philips booth right next to their demo stations for the CDTV. Before the grand opening of the convention, the folks who worked at Philips came over and started playing on our technology. They were so impressed they would nod their heads and say things like, "Impossible! How? Wow, it blows away our product in speed! It's expensive but cool," etc. We smiled, knowing that the technology had proven itself and was now being judged by its peers. Before we could get a big head over it, our elation turned into a life lesson. When the convention opened, we attracted more of a crowd to our exhibit than Philips did to theirs—in the same booth! This made the top executive at Philips uneasy. He decided he wanted us to shut down our demo so we would not detract from his product. When they told us, I defended our right to stay and challenged him to look at the situation differently. I suggest that we tell people that our product is the ultra-high end of interactive capabilities and is expensive, and the CDTV is the consumer equivalent and affordable. By accepting this deal, he allowed more people to see his and our products.

A huge marketing lesson I learned is to find a way to put your pitch front and center. My small team directed people to the Philips products, and we got to stay in the booth, thereby converting a problem into a marketing solution for everyone involved.

When we created the interactive Playable Television, no one believed it was possible because they could not envision it and did not think technology could do it back in the 90s. The owner and founder of NewTek, Tim Jenison, who created the Video Toaster, was not convinced we had turned his video-switching product into an interactive self-switching, essentially a robot. For techies, that meant we had automated his video switcher and put a human touchscreen interface in front of the switcher video with virtual transparent buttons overlaid. When these buttons were pressed, they would create interactive live outcomes.

To get more Video Toasters, I had to demo my product for Tim Jenison at NewTek's headquarters in Topeka, Kansas. Jason and I flew out with our prototype. When we arrived at NewTek's corporate office, my eventual mentor Tim took our prototype into a small side room and said he would privately "test" the unit to see if it was stable doing real-time functions. I realized later that he was pushing it to the limit to learn its capabilities. Tim was intent on finding the breaking point. He was gone for some time but came to where we were waiting in a large conference room. He didn't say a word but grabbed the bowl of choc-olate kisses in the middle of the table. Thirty minutes later, he returned with an empty bowl, ignoring us as he directed a staff member to refill it. It's like we weren't even in the room. He got his refill and returned to the side room to continue his "test drive" of our machine. All kinds of things were rolling around in my head. I had deemed it a torture room, the death of our Playable Technologies. I was sure he would complete his goal of crashing our system, and we were done. What goes through your head while you wait for your fate is amazing! Finally, he emerges from the room with the biggest smile, "I tried as hard as I could to crash her, but I could not. Congratulations, this is cool technology and a new application for the Video Toaster. You did it, guys!"

That is how technology is created. You dare to try stuff no one has done before and take risks, knowing you can always fail. But succeeding is beyond amazing, and that feeling you get when a tech titan like Tim says, "You did it!" is why I create technology. I don't control how I get projects—that appears to be the work of a higher power—but rather how I create and make my clients happy. I am proud to be a technologist and do what I do because I can. That is really what I discovered is the singular truth of why I was made to do what I do, and everything about me was designed perfectly for my purpose. When you understand yourself and your gifts to the world, you become responsible for making good on them.

## From ANC to the Airwaves

My life's journey seemed always to intertwine technology and entertainment unexpectedly. After my success with Disney's Wide World of Sports, a contact introduced me to Jerry Cifarelli, owner of ANC Sports, a company that installed LED video boards in sports stadiums.

It was around 1998 when Jerry invited me to fly to the Pepsi-Cola Center, a moment etched in history. Inside, I beheld a 360-degree video sign, an extraordinary creation marred by fragmentation. It was seventy disjointed displays, each separate and solitary, unable to form a harmonious symphony of visuals. These rectangular canvases had limits, only capable of showcasing snippets of information.

A spark of inspiration struck me—*Could I transform this disjointed ensemble into a contiguous display? Could I transform it into a true 360-degree experience?* The answer was a resounding yes. I embarked on a quest to revolutionize LED displays, and Jerry enthusiastically found the capital to make it all happen. Jerry taught me business and explained the risks he took to get the capital. I was motivated by reality for us to succeed. Nothing would have happened without Jerry getting huge loans in the millions to cover production research and development. Ideas need money. With trips to China to revise the video board hardware and my team working on the software, we finally unveiled to the world, the first 360-degree video board, a marvel that found its home in Ottawa and Washington, DC, along with Canada's Air Canada Centre. My role with ANC started as a consultant, and today, I am its Chief Technology Officer.

Around 2020, another chapter emerged from a rather unexpected angle. I have a passion for music that has always been a hobby. With my third album freshly out, Jerry urged me to share my music on air at Ocean 98.1 on the *Rude Awakening Show*. I sent Bulldog my third music album creation, a musical venture I held dear. To my surprise, he embraced it enthusiastically, and that's how I found myself on the radio.

What followed was an unexpected twist. They aimed to quiz me on sports statistics, a domain in which I wasn't well versed. Little did they know, their "roasting" was tipped off, so I had meticulously gathered insights from experts to fortify my arsenal. Live on the air, I turned the tables on them, posing questions they couldn't answer, and I delivered answers to their questions with finesse. The result? It was a thrilling radio show—a dance of intellect and wit that had Bulldog proclaiming it among his top shows to me. This serendipitous encounter with Bulldog set the wheels in motion for something greater.

Jerry, in his ever-wise capacity, decided it was time for a new venture. This was around the onset of the COVID-19 era, and he was no longer my direct superior after the company transitioned. Yet, his final directive resonated deeply—pitch Bulldog a technology show. Now, you must understand I wasn't versed in radio nor did I possess the confidence to believe I could secure a slot. But Jerry's insistence, borne of an undeniable belief in me, spurred me on.

I woke the next day, a whirlwind of thoughts and ideas engulfing me. "Tech Byte—the edge in everyday technology" was the phrase that echoed in my mind, a technology-oriented show that would encapsulate everyday innovation. It was this idea that I nervously presented to Bulldog, unsure of his reaction. The response was overwhelmingly positive—Wednesday at 8:00 for me, 9:00 for him—the time slot secured.

As time passed, my radio journey grew—to bantering with the Mayor of Ocean City. But amidst all this, I recognized that, sometimes, inspiration strikes from the most unexpected quarters. This, my friend, is the tale of my technological odyssey intertwined with radio, a journey that defied conventions and where the extraordinary emerged from the ordinary.

I have spent decades discovering, developing, and exploiting new technologies, from broadcast television to 3D graphics, music composition to distribution, and now, the largest, most dynamic sports and event experiences in the world—digital experiences such as The World Trade Center, Penn Station, NBC's Tonight Show and SNL, and the transforming scoreboard of the 76ers—all controlled by my software built by an amazing team of people. My broad array of technological encounters has uniquely equipped me for the perspective I have shared in this first book, of many, to help humans live as humans and enjoy technology in a healthy way for generations to come.

Milton Keynes UK
Ingram Content Group UK Ltd.
UKHW020742280424
441851UK00013B/168/J

9 798218 970376